Middle School Algebra for the Common Core

Middle School Algebra for the Common Core

 Published in the United States by LearningExpress, New York.

Cataloging-in-Publication Data is on file with the Library of Congress.

Printed in the United States of America

9 8 7 6 5 4 3 2 1

For more information or to place an order, contact LearningExpress at:
80 Broad Street
4th Floor
New York, NY 10004

Or visit us at:
www.learningexpress.com

Contents

Contents

Contents

About the Author

Kimberly Stafford majored in mathematics and education at Colgate University in upstate New York. She taught math, science, and English in Japan, Virginia, and Oregon before settling in Los Angeles. Stafford began her work in Southern California as an educator in the classroom but soon decided to launch her own private tutoring business so she could individualize her math instruction. She believes that a solid foundation in math empowers people by enabling them to make the best work, consumer, and personal decisions. Stafford is unfazed by the ubiquitous student gripe, "When am I going to use this in real life?" She stresses that the mastery of math concepts that are less applicable to everyday life helps teach a critical skill—problem solving. The very ability to apply a set of tools to solve new and complex problems is an invaluable skill in both the workforce and personal life. Stafford believes that mathematics is a beautiful arena for developing organized systems of thinking, clear and supported rationale, and effective problem solving.

Getting to Know the Common Core State Standards

Does even thinking about the Common Core State Standards make your palms sweat a bit and your heart rate elevate slightly? Well, you are not alone. New requirements and standards are always a little nerve-wracking—and not just for teachers, but also for parents, students, and school districts alike. There's been a lot of buzz about the Common Core State Standards along with a lot of misinformation. (Moving forward, we'll occasionally use the abbreviation *CCSS* for the Common Core State Standards.) People say that the CCSS were developed without input from teachers, that the standards are a narrow "script" that teachers cannot deviate from, and that the standards keep kids from connecting concepts to the real world, but all of these claims are untrue. The fear of change drives much of the misinformation out there, so let's take a closer look at the development, rationale, and goals of the Common Core State Standards.

Much of the backlash against the Common Core State Standards is fueled by misinformation regarding who developed the standards, why they were developed, how they are implemented in schools, and what they mean for your school district's teachers and children.

Who?

The idea of having a body of common standards that would cross state lines was introduced by the National Governors' Association Center for Best Practices (NGA Center), along with the Council of Chief State School Officers (CCSSO). Contrary to what some people believe, the standards were not designed by the federal government, but were instead developed by a diverse group of educators and experts from all around the country. Along with teachers, scholars, assessment developers, and parents, the following well-respected professional organizations provided feedback on the standards:

- National Education Association (NEA)
- American Federation of Teachers (AFT)
- National Council of Teachers of Mathematics (NCTM)
- National Council of Teachers of English (NCTE)

The CCSS were developed by well-respected professional organizations along with the help of teachers, scholars, assessment developers, and parents.

Why?

The United States educational system has not kept pace with the skills gained by students in many other industrialized countries. Although the United States once ranked first for high school graduation, it has slipped to 22nd among the top 27 industrialized countries, and college retention rates are also plummeting. For students who successfully make it into the halls of higher education, university instructors and officials are noticing that their preparedness levels are noticeably declining. The Common Core State Standards aim to raise the bar for the material presented to students in K–12. The CCSS are not only *aligned across state lines*, but they are also *internationally benchmarked* to help students gain the skills they need to compete in the global market. The desired outcome of the CCSS is to give students the applied skills they need for college, work, and life.

What?

The Common Core State Standards are a collection of learning goals that outline the skills students should gain at each grade level. The standards give students and parents a clear understanding of the knowledge needed to be college- and career-ready upon graduating high school. The standards cover skills in mathematics and English language arts/literacy from kindergarten through 12th grade. One of the main aspects of the mathematics standards is a shift away from a body of skills that is "a mile wide and an inch deep." Instead of rapidly moving from one concept to the next, the standards present a more focused set of skills that require students to engage in higher-level reasoning and problem solving. The standards are:

1. Based on thorough research and evidence
2. Able to be clearly understood by educators, parents, and students
3. In alignment with the expectations of colleges and employers
4. Rigorous in content and require students to apply higher-level reasoning and problem-solving skills
5. A dynamic by-product of the state standards that have been most effective, and a revision of shortfalls in previous state standards
6. Intended to prepare students to succeed in the global workforce by aligning with the educational standards used by top-performing countries

The CCSS are aligned across state lines and are internationally benchmarked to help students gain the skills they need to compete on the global market. The mathematics CCSS move away from a curriculum that is "a mile wide and an inch deep." A more focused curriculum helps students develop the most essential skills that encourage higher-level reasoning and problem solving.

How?

The Common Core State Standards are not to be thought of as a narrow curriculum that cannot be deviated from; the standards are a broad collection of skills that do not include each and every skill students should learn. How school districts and teachers present the standards is determined by how they choose to design their curricula. There are no strict lesson plans to teach from and educators are encouraged to tailor their teaching lessons to best meet the individual needs of their students.

When?

State education standards were popular by the early 1990s. It wasn't too long before all states had their own independent learning standards, including their own ideas of what defined "proficiency." The lack of a unified common body of standards meant that kids who were excelling at the standards established by one state might not have been meeting standards set by a different state. Educators saw a need for a cross-state body of learning standards, and work on the Common Core State Standards began in 2009. The standards were introduced to states for adoption in 2010 and states are independently implementing the CCSS.

Where?

The Common Core State Standards are not a federal mandate forced on schools. The standards have been voluntarily adopted by 43 states. Each state that has adopted the CCSS works to deliver the same general content to each grade level. This benefits students, because a student finishing 7th grade in Idaho and entering 8th grade in Florida the following school year will be able to be at the same level as her peers. Previously, students changing school districts across state lines were in jeopardy of having their education compromised because of gaps in content.

The lack of a unified body of standards meant that kids who were excelling in one state might not been able to meet the standards set by a different state. Previously, students changing school districts across state lines were in jeopardy of having their educations compromised because of gaps in content.

Visit this site to learn more about the CCSS:

http://www.corestandards.org/about-the-standards/

Algebra All Around Us!

Now that you know a little more about the rationale behind the Common Core State Standards, they're a little less scary, right? Maybe you think that it even sounds like a good idea to have a unified set of sequential skills that will help students prepare for the college classroom or a job in the workforce. Now that we've covered some information on the standards in general, let's take a closer look at what middle school math standards relate to algebra and narrow the focus of what we will cover in this book. The math standards in grades 6–7 are broken into the following six topics (the first five topics are covered in 6th and 7th grade and the last topic is added in 8th grade):

- Ratios & Proportional Relationships
- The Number System
- Expressions & Equations
- Geometry
- Statistics & Probability
- Functions

The subtopics of *Geometry* and *Statistics & Probability* are both comprehensive fields of math that encompass wide bodies of skills. Although there is some overlap with algebra in these topics, *Geometry* and *Statistics & Probability* both deserve their own book to properly present and discuss the Common Core State Standards relating to them. Therefore, we will not cover *Geometry* and *Statistics & Probability* CCSS in this book. Similarly, the subtopic *The Number System* will not be a focus in this book since the standards in this arena focus more on *arithmetic* (the laws and principles of the basic operations of addition, subtraction, multiplication, and division on numerical expressions) than on *algebra* (using variables to model mathematical relationships through variable expressions).

Algebra is an invaluable field of math. With algebra, models can be created that represent real-world situations. These models can be solved to answer specific questions or can be used to recognize trends and make predictions. Algebra is the departing point from the concrete mathematical operations of arithmetic and the entry point into more abstract ways of thinking through the use of variables. Making this transition helps students develop solid reasoning skills that can be used to solve a range of mathematical and scientific problems inside and outside the classroom.

Although success in algebra is a prerequisite for success in higher-level math classes in high school and in obtaining a high school diploma, algebra is also a vital skill outside of high school campuses. For students who decide to head directly into the workforce from high school, studies show that a solid background in algebra literally pays off. A study performed by the ACT found that occupational fields that don't require a college degree, but still offer large enough wages to support a family of four, require the same math skills that college freshmen need to have in order to be successful in college! Did you know that fields like plumbing, electronics, and upholstery can pay great wages without a college degree? Therefore, in order to have the opportunity to excel in fields that offer solid wages without a college education, it is important to develop solid algebra skills. So, whether you can't wait to graduate high school so that you can get a job or pursue a college education, algebra success should be part of your game plan!

The CCSS for Middle School Algebra

Remember, an important aspect of the Common Core State Standards is the increased focus it brings to a more restricted list of topics. There has been a departure from the idea to teach a greater number of concepts with little depth and rote memorization. Instead, the CCSS puts forth a refined set of concepts with increased vigor. The CCSS challenges students to have a more grounded understanding of the origins and applications of theorems, algorithms, and formulas; students will be asked to apply mathematical ways of thinking to solving real-world challenges.

The following is a brief breakdown of the algebraic skills students will cover each year. The 6th grade standards introduce students to rates and ratios and algebraic modeling. They don't do a lot of rigorous work with variables just yet, but students will certainly learn to apply the **properties of operations** on variables. Specifically, the 6th grade algebraic skills will focus on:

- Understanding ratios, rates, and unit rate in tables, equations, and graphs
- Investigating percents as special types of ratios

- Working with exponents and using the order of operations to simplify numerical expressions
- Understanding what variables are and representing situations with variable expressions
- Evaluating algebraic expressions and solving one-step algebraic equations

In 7th grade, students will begin do to some heavy lifting with algebraic expressions after they have extended their knowledge of ratios to proportions. The 7th grade algebraic skills will focus on:

- Using **ratios** and **proportions** to construct algebraic models, tables, and graphs
- Solving **percent** problems with proportions
- **Real-world problem solving** with multi-step word problems and inequalities
- Solving two-step algebraic equations

Students should arrive to 8th grade fearless of variables. They will be refining their skills with exponents and algebraic equations and will then move on to systems of equations and functions. Before heading to high school students will:

- Learn and apply the laws of exponents and become proficient in scientific notation
- Learn how to write equations in slope-intercept form
- Apply problem-solving skills to real-world, multi-step linear equations and systems of equations
- Understand what functions are, how to recognize them, and how to analyze graphs of functions

How to Use This Book

We know you're revved up now and excited to get down with algebra as soon as possible! The first 10 lessons cover 6th grade standards, Lessons

11–18 cover 7th grade standards, and Lessons 19 through 28 cover 8th grade standards. Each lesson introduction will present what standard or standards will be highlighted in that lesson. Even if you are beyond 6th grade, it is still a good idea for you to begin with Lesson 1 since the new standards use language and terms that may not have been used in your 6th or 7th grade classrooms.

While some parents or students find this new language to be a frustrating shift, its aim is actually to use language that applies more to the real world. Sure, they could insist on just sticking with the same language that was used in your great-grandfather's classroom, but that doesn't really make sense if those terms are too abstract to apply to our lives. Parents may notice that terms like "starting point" have replaced "y-intercept" and "rate of change" has replaced "slope." It just takes a little getting used to, but in the long run, these new ways of thinking about algebraic concepts should make the skills students learn easier to apply to their everyday lives.

This book is structured in logical sequences, so that each lesson builds on the previous lessons. Therefore, it is a good idea to move forward through the book from the beginning, to make sure that you fully understand each lesson's material before proceeding to the next lesson. Each lesson has examples, illustrations, ERROR ALERTS, and boxed highlights that you want to make sure you fully comprehend. Each Lesson has two to four sets of practice problems along with answers to those problems in the back of each lesson. You will notice that the practice sets are not long lists of repetitive problems. Life does not throw the same pitches every day; instead, we constantly need to adapt to new types of situations and problems that may arise. Therefore, we try to challenge you with questions that require you to have an in-depth understanding of the material that was presented, and sometimes you are given questions that will ask you to extend your knowledge to consider topics that you will be learning next. We do not give you the hardest questions possible, but also not the easiest questions—we have aimed for a middle ground.

The answers we provide to the questions in each lesson are thorough, so the more you put into your work on these questions the more you will get out of this book. By investing focused study time on each lesson, you should be able to develop your algebra skills enough so that you feel confident in applying them to real-world problem solving.

For more information about the mathematics standards visit
www.corestandards.org/Math/

Middle School Algebra for the Common Core

1

Introducing Ratios and Unit Rates

STANDARD PREVIEW

In this lesson we will cover **Standards 6.RP.A.1** and **6.RP.A.2**. You will be introduced to ratios, which are comparisons of two quantities. You will use ratios to find unit rate. You will also practice finding unit rates with fractions, seeing that unit rates with fractions require the same steps as unit rates with whole numbers.

Understanding Ratios

A **ratio** is a way of using a fraction to compare two numbers. We use ratios every day, often without even being aware of it. When your teacher puts gas in his car, he's probably wondering how many gallons *per* mile his car is getting. When your bus driver checks the speedometer she's looking to see how many miles *per* hour she's going. When your parents look at the ingredients on a box of cereal they might be looking for how many grams of sugar it has *per* serving. The list of ratios that are used by people daily goes on and on! You may have noticed above that the word *per* occurred in each example—per is actually a great indicator that two pieces of information can be expressed as a ratio.

The Language of Ratios

Simply stated, a ratio is a fraction. What makes ratios special is the way mathematicians use them to describe and compare things in a real-world context. One example that is commonly used to introduce ratios is the number of boys and girls in a classroom. Suppose there are 25 students in a class: 10 are boys and 15 are girls. (The expression *for every* is often used with ratios, so this information might be presented as, "For every 10 boys in the class there are 15 girls.") There are several different ways of using ratios to describe this class. The ratio of boys to girls is 10 to 15. This can also be written as 10:15 or $\frac{10}{15}$, although for reasons we will see in a moment, the fraction form does not make much sense in this type of ratio when we are comparing a part to another part. Some other ratios that model the class include:

- The ratio of girls to total students is 15 to 25
- The ratio of boys to total students is 10:25
- Boys comprise $\frac{10}{25}$ of the class

Just like with regular fractions, it is best to express a ratio in simplest form. If the ratio of boys to girls is 10 to 15, that should be reduced to an equivalent ratio of 2 to 3. In other words, for every 2 boys in the class, there are 3 girls. Notice that this does not mean there are *only* 2 boys and 3 girls in a class—the class could be divided into groups of 5, with 2 boys and 3 girls in *every* group. Therefore, if the ratio of boys to girls in a class is 2:3, we would describe this as meaning that *for every 2 boys in the class there are 3 girls*.

Two Types of Ratios

There were three quantities used in the previous ratios: boys, girls, and total students. *Boys* are a part, and *girls* are a part, but *total students* represents the whole. There are two different types of ratios that are often useful for comparing items:

- **Part-to-part ratios** compare the sizes of two different parts of the whole. In this case, it usually doesn't make sense to use the fraction form, since the denominator does not represent the whole of anything. The ratio of *boys* to *girls* and the ratio of *girls* to *boys* are examples of part-to-part ratios.

- **Part-to-whole ratios** compare the size of one part to the size of the whole, just like a typical fraction where the numerator is the part and the denominator is the whole. The ratio of *boys* to *total students* and the ratio of *girls* to *total students* are examples of part-to-whole ratios.

Practice 1

1. What is a part-to-part ratio and what is a part-to-whole ratio? Which one of these makes more sense to represent as a fraction?

2. In the first hour open, Prima's Donut shop sells 20 cups of coffee and 36 donuts. Represent this information using one part-to-part ratio and two different part-to-whole ratios in simplest form.

3. Use your simplest form answers from question 2 to explain what each ratio means in real-world sales for the donut shop.

4. For every seven minutes Babs runs she stretches for two minutes after her workout. Write three different ratios that could represent Babs' running to stretching ratio.

5. Using the ratio from question 4, if Babs runs for 35 minutes, how long will she stretch for? If Babs stretches for 20 minutes, how long did she run for?

Use the following to answer questions 6 and 7:

Too Many Terriers Animal Shelter takes in homeless dogs and cats. The ratio of dogs to the total number of animals in the shelter on a given day is 3:10.

6. What does the given part-to-whole ratio mean? Describe the situation above using a part-to-part ratio of dogs to cats.

7. If there are 28 cats in the shelter, how many dogs are in the shelter?

Rate and Unit Rate

Like ratios, we use rate every day without even realizing it. A **rate** is a special type of ratio that compares two measurements that have different units, like lawns and days: *Polina mowed 12 lawns in three days.* A **unit rate** is a rate that compares the first type of measurement to just 1 unit of the second type of measurement. The rate of *12 lawns in three days* can also be expressed as a unit rate: *Polina mowed four lawns per day. Rates* are often presented in fractional form and a *unit rate* is a fraction with a denominator of one. Unit rates are presented most commonly as a phrase in the *per unit* language. We see unit rate all around us: words per minute, price per pound, and persons per square mile are all examples of unit rates. Although unit rate is used in many different contexts, we are going to focus here on examples of unit rate in speed, price, and population density.

Unit rate shows how many units of one type of quantity correspond to *just one* unit of a second type of quantity. It can be expressed as a rate with a denominator of one but is more often presented in *per unit* phrasing.

Speed

It took Selma one hour and 45 minutes to walk a four-mile trail. How many miles per hour did she walk? Round your answer to the nearest tenth.

To find how many miles *per hour* Selma walked, we write our fraction with the number of miles as the numerator and the number of hours as the denominator. (The per *hour* tips us off that *hours* go in the denominator.):

$$\frac{4 \text{ miles}}{1.75 \text{ hours}}$$

Since we want to see how many miles Selma walked in just one hour, divide the numerator and denominator by 1.75 to reduce the fraction:

$$\frac{4 \text{ miles}}{1.75 \text{ hours}}\left(\frac{\div 1.75}{\div 1.75}\right) = \frac{2.28 \text{ miles}}{1 \text{ hour}}$$

Rounded to the nearest tenth, Selma walked 2.3 miles per hour.

What if, instead of wondering how many miles Selma walked *per hour*, you wanted to see how many minutes it took Selma *per mile*? Here, the *per* is tipping us to put the *miles* in the denominator, so let's use the same technique:

$$\frac{1.75 \text{ hours}}{4 \text{ miles}}$$

Now divide the numerator and denominator by four to reduce the fraction so that it is in terms of just one mile:

$$\frac{1.75 \text{ hours}}{4 \text{ miles}}\left(\frac{\div 4}{\div 4}\right) = \frac{0.44 \text{ hours}}{1 \text{ mile}}$$

So, it took Selma 0.44 hours to walk 1 mile. (We'll talk more about this later, but 0.44 hours is less than half an hour, so make sure you don't mistake this for 44 minutes! To calculate what $\frac{44}{100}$ of an hour is, multiply 0.44 by 60 since there are 60 minutes in an hour: 26.4 minutes.)

The quantity that follows the word *per* goes into the denominator and you are finding the unit rate of that quantity. Example: Alana earns $20 *per* two hours: $\frac{\$20}{2 \text{ hours}} = \frac{\$10}{1 \text{ hour}}$ **= $10 per hour.**

Prices

Rick's Market is selling 12 lb. turkeys for $19.50 each, and Mike's Meats is selling 15 lb. turkeys for $23.85. Which store offers the better price per pound of turkey?

To find which store offers a better price per pound of turkey, we need to calculate the unit rate for turkey at each store. The price at Rick's Market is $19.50 per 12 pounds, which can be written as:

$$\frac{\$19.50}{12 \text{ lbs.}}\left(\frac{\div 12}{\div 12}\right) = \frac{\$1.63}{1 \text{ lb.}}$$

When simplified, we find that the rate at Rick's Market is $1.63 per pound.

The price per pound at Mike's Meats is $23.85 per 15 pounds, which can be written as

$$\frac{\$23.85}{15\text{ lbs.}}\left(\frac{\div 15}{\div 15}\right) = \frac{\$1.59}{1\text{ lb.}}$$

The rate at Mike's Meats is therefore $1.59 per pound. Mike's Meats offers a better price per pound of turkey, but only by four cents!

Density

An estimated 392,880 people live within 58 square miles in Minneapolis, Minnesota. Approximately 3.82 million people live within 503 square miles in Los Angeles, California. What is the difference of people per square mile between these two cities? Round to the nearest whole number.

To find the persons per square mile in Minneapolis, we need to first write our fraction $\frac{392{,}880\text{ people}}{58\text{ square miles}}$. When simplified, the number of people per square mile in Minneapolis is 6,774.

To find the people per square mile in Los Angeles, CA, we write another fraction: $\frac{3{,}820{,}000\text{ people}}{503\text{ square miles}}$. When simplified, the number of people per square mile in Los Angeles is 7,594.

To find the difference of people per square mile between Minneapolis and Los Angeles, we subtract 6,774 from 7,594.

$7{,}594 - 6{,}774 = 820$ people per square mile

Unit Rate with Fractions

Notice that even if the given information contains fractions or mixed numbers, the method is still the same. Also, it is important to remember that dividing by a fraction is the same thing as multiplying by its reciprocal.

If it takes Joanie $\frac{1}{4}$ of an hour to walk $\frac{3}{4}$ of a mile, find her unit speed per mile.

Since you are looking for the unit speed per mile, put miles in the denominator and then change the division of a fraction to multiplication by multiplying the numerator by the reciprocal of the denominator:

$$\frac{\frac{1}{4}\text{ hour}}{\frac{3}{4}\text{ mile}} = \frac{1}{4} \times \frac{4}{3} = \frac{1}{3}\text{ hour} = 20\text{ minutes per mile}$$

Practice 2

1. Anita is knotting silk necklaces for an order placed by a large department store. On Monday she made 28 necklaces in eight hours. How many necklaces does Anita make in an hour?

2. Now use the information from question 1 to determine how long it takes Anita to make one necklace. What is her unit time per necklace?

3. Pete, Celia, and Lauren are planning a road trip to Zion National Park. Pete's car has a 12-gallon tank that gets 288 miles. Celia's car has a 15-gallon tank that gets 330 miles. Lauren's car has an 18-gallon tank that gets 378 miles. Help them determine who should drive by calculating how many miles per gallon each person's car gets.

4. A jet can fly a distance of 2,430 miles in $4\frac{1}{2}$ hours. Find the jet's average speed in miles per hour.

5. If the jet in question 4 travels at the same average speed throughout its flight, how far will it travel in $13\frac{1}{2}$ hours?

6. Eliseo made $90.00 babysitting for 12 hours. Find Eliseo's babysitting hourly rate and determine how much he will earn if he works Monday through Friday, eight hours a day.

7. It takes Ann $\frac{1}{5}$ of an hour to make $\frac{1}{12}$ of a scarf she is knitting. How long does it take her to knit one scarf?

8. Use the information from question 7 to determine what fraction of a scarf Ann can knit every hour.

Answers

Practice 1

1. A part-to-part ratio is a comparison of two different quantities that are both parts of a whole. A part-to-whole ratio is a comparison of a part of something to the total number of items that comprise the whole. It makes sense to represent part-to-whole ratios as fractions, but not part-to-part ratios.
2. Cups of coffee to donuts: $\frac{20}{36} = \frac{5}{9}$; cups of coffee to total sales: $\frac{20}{56} = \frac{5}{14}$; and donuts to total sales: $\frac{36}{56} = \frac{9}{14}$.
3. $\frac{5}{9}$ means that for every nine cups of coffee sold, five donuts were sold. $\frac{5}{14}$ means that for every 14 items sold, five of them were cups of coffee. $\frac{9}{14}$ means that for every 14 items sold, nine of them donuts.
4. 7:2, 14:4, 21:6, 28:8, and any ratio that reduces to 7:2 is an example of Babs' running to stretching ratio.
5. If Babs runs for 35 minutes, she will stretch for 10 minutes because 35:10 is the same ratio as 7:2. If Babs stretches for 20 minutes, that means she is performing 10 sets of running and stretching. 10 sets of 7 is 70, so Babs ran for 70 minutes. Nice going!
6. The part-to-whole ratio of 3:10 means that for every 10 animals at the shelter, three of them are dogs. The ratio of dogs to cats is 3:7, so for every three dogs there are seven cats.
7. If there are 28 cats that means there are four sets of 3:7 dogs to cats in the shelter. Four sets of three dogs means there are 12 dogs.

Practice 2

1. To find the unit rate per hour: $\frac{28 \text{ necklaces}}{8 \text{ hours}} = 3.5$ necklaces per hour.
2. To find the unit rate per necklace: $\frac{8 \text{ hours}}{28 \text{ necklaces}} \approx \frac{0.29 \text{ hours}}{1 \text{ necklace}}$. Changing 0.29 hours to minutes by multiplying by 60 gives a unit rate of approximately 17 minutes per necklace.
3. Pete's: $\frac{288 \text{ miles}}{12 \text{ gallons}} = 24$ miles per gallon. Celia's: $\frac{330 \text{ miles}}{15 \text{ gallons}} = 22$ miles per gallon. Lauren's: $\frac{378 \text{ miles}}{18 \text{ gallons}} = 21$ miles per gallon. Looks like Pete should drive!
4. $\frac{2{,}430 \text{ miles}}{4.5 \text{ hours}} = 540$ miles per hour.
5. Since the jet's speed is 540 miles per hour, multiply $13\frac{1}{2}$ hours by 540 miles to get 7,290 miles.
6. $\frac{\$90}{12 \text{ hours}} = \$7.50$ per hour. A Monday–Friday workweek of eight hours per day contains 40 hours so multiply 40 by Eliseo's hourly rate: 40 hours $\times$ \$7.50 = \$300.

7. Since we are looking for Ann's *per scarf* speed, put the scarf in the denominator and time in the numerator:

$\dfrac{\frac{1}{5}\text{ hour}}{\frac{1}{12}\text{ scarf}} = \frac{1}{5} \times \frac{12}{1} = \frac{12}{5}$ hour = $2\frac{2}{5}$ hours. Divide 60 minutes by five to find that $\frac{1}{5}$ equals 12 minutes. Therefore, $2\frac{2}{5}$ hours = two hours and 24 minutes.

8. Since we are looking for Ann's *per hour* rate, put the time in the denominator of the fraction and scarf in the numerator:

$\dfrac{\frac{1}{12}\text{ scarf}}{\frac{1}{5}\text{ hour}} = \frac{1}{12} \times \frac{5}{1} = \frac{5}{12}$ of a scarf per hour.

Therefore, since $\frac{4}{12}$ is the same as $\frac{1}{3}$, and Ann knits $\frac{5}{12}$ of a scarf an hour, we conclude that she knits a little more than one-third of a scarf every hour.

2

Rates, Tables, Conversions, and the Coordinate Plane

STANDARD PREVIEW

In this lesson we will cover **Standards 6.RP.A.3.A**, **6.RP.A.3.B**, and **6.RP.A.3.D**. You will learn how to use ratio and rate reasoning to solve real-world unit rate problems and how to find the missing values in tables. We will also use ratio tables to convert from one unit of measurement to another. Lastly, we will plot pairs of values on the coordinate plane.

Representing Ratios in Tables

A ratio table is a handy tool for solving ratio problems. Ratio tables clearly display the patterns that emerge when ratios are used to solve problems. Mastering ratio tables is an important tool in deepening your understanding of rates and ratios.

Ratios in Recipes

A great example for a ratio table is to think about a drink mix. To make her summer fruit punch, Jasmine mixes five cups of grape juice with two cups of cranberry juice. Jasmine made the following table to figure out the ingredient ratios when she is mixing juice for large parties by multiplying the original ratio of 5:2 by factors of 2, 3, 4, and 5:

Grape Juice	5 cups	10 cups	15 cups	20 cups	25 cups
Cranberry Juice	2 cups	4 cups	6 cups	8 cups	10 cups

A ratio of 5:2 is equivalent to a ratio of 15:6 or 20:8. All of these are part-to-part ratios. If Jasmine wants to know the total quantity of fruit punch she will be making, she can add another row to her table that displays the total number of cups of punch:

Grape Juice	5 cups	10 cups	15 cups	20 cups	25 cups
Cranberry Juice	2 cups	4 cups	6 cups	8 cups	10 cups
Total Fruit Punch	7 cups	14 cups	21 cups	28 cups	35 cups

All the relationships between the columns in a ratio table are always based on multiplication. This means that to create new numbers in one column we can multiply both numbers in the original column by the same factor:

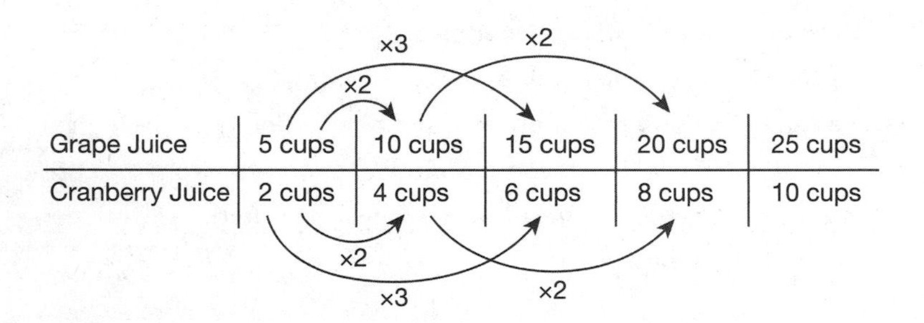

Many students have the misconception that this relationship should be additive. That is, that you should add the same number to go from one column to the next. Remember that ratios are fractions and that in order to create equivalent fractions multiplication is used and not addition. Look how multiplication can be used in the following table to move from one column's ratio to another column's ratio. (The table shows the total sales income for several different numbers of tickets sold.)

# of Tickets Sold	1	2	10	50	100	1,000
Total Sales Income	$6	$12	$60	$300	$600	$6,000

In this example, we multiply the number in a column by any number to create new numbers in another column—for example, multiplying the second column by 25 can help us arrive at the fourth column:

2 tickets × 25 = 50 tickets
$12 × 25 = $300

When working with ratio tables, use multiplication to create equivalent ratios, not addition. (Division can also be used to reduce ratios.)

Finding Missing Values

You can make a ratio table out of any ratio. Suppose you are buying personal pizzas for a birthday party. You have a coupon to get 3 pizzas for $10.

Pizzas	3	6	9	12
Cost	$10			

You can use the multiplicative nature of fractions and ratios to find the costs for six, nine, and 12 pizzas. You don't need to figure out the cost of an individual pizza. Since the number 6 is twice as much as three, six pizzas would cost $20 (2 × $10). Similarly, nine pizzas would cost $30 and 12 pizzas would cost $40. Notice in the table below how each new value was calculated:

Pizzas	3	(3 × 2) 6	(3 × 3) 9	(3 × 4) 12
Cost	$10	($10 × 2) $20	($10 × 3) $30	($10 × 4) $40

Determining the Multiplication Relationship

Making a rate table may not seem like a very complex idea, but when the numbers become more challenging, or the context becomes more abstract, it can get tricky. Sometimes the multiplication relationship will not be so apparent and you will need to use division to find it:

Example: *On a road trip, it took $2\frac{1}{2}$ hours to drive the first 150 miles. Estimate the time it will take to drive 300 miles, 450 miles, and 100 miles.*

Solution: A ratio table can be very helpful here for estimating driving times for the rest of the trip.

Distance	150 miles	300 miles	450 miles	100 miles
Time	2.5 hours			

Since $150 \times 2 = 300$, we can multiply 2.5 hours by two to estimate the driving time for 300 miles: $2.5 \times 2 = 5$ hours. The same technique works for 450 miles: $150 \times 3 = 450$ miles, so 2.5 hours $\times 3 = 7.5$ hours.

But suppose there are only 100 miles left on the trip. It would be convenient if we knew what to multiply 150 by to get 100 because then it would be clear what the multiplication relationship is between the given ratio and the desired ratio. Let's determine what factor we need to multiply 150 miles by in order to yield 100 miles, and then we can use that factor to determine the number of hours:

$$150 \times ? = 100$$

Since the opposite of multiplication is division, divide 100 by 150 in order to determine what the "?" is in the equation:

$$100 \div 150 = \frac{100}{150} = \frac{2}{3}$$

Now we know $150 \times \frac{2}{3} = 100$, so if we also multiply 2.5 hours by $\frac{2}{3}$, we'll get the approximate driving time for 100 miles. We will write $2\frac{1}{2}$ hours as $\frac{5}{2}$ hours so that we can multiply by fractions:

$$\frac{5}{2} \times \frac{2}{3} = \frac{5}{3} = 1\frac{2}{3}$$

It will take $1\frac{2}{3}$ hours, or one hour 40 minutes (since $\frac{2}{3}$ of 60 minutes is 40 minutes), to drive the last 100 miles. Fill in the table with the correct values for each distance:

Distance	150 miles	300 miles	450 miles	100 miles
Time	2.5 hours	5 hours	7.5 hours	1 hour, 40 min

Rate Reasoning

Now that you are comfortable finding the ratio of two quantities, determining their unit rate, and finding missing values in tables, let's look at how to apply ratio tables to solve a problem with unit pricing.

Example: *Jean is buying organic goji berries that cost $8.59 for six ounces. If she has $50 to spend on her goji berries for the month, how many ounces can she buy?*

Solution: Put the given information in a ratio table in order to determine how to find the multiplication relationship between the given rate and the information you are trying to find:

Goji Berry Price	$8.59	$50
Goji Berry Ounce	6 ounces	? ounces

In this case, divide $50 by $8.59 in order to identify the multiplication relationship:

$50 ÷ $8.59 = 5.82.

Multiply six ounces by 5.82 to see how many ounces of berries Jean can buy: 6 × 5.82 = 34.92 ounces, which we'll round to 35.

Goji Berry Price	$8.59 × 5.82 =	$50
Goji Berry Ounce	6 oz × 5.82 =	35 ounces

Practice 1

1. Skott's Skin Care sells a face serum that costs $138 for 3.4 ounces. They also offer a travel-sized bottle that costs $57.50. If the travel size is the same cost per ounce, how many ounces are in the smaller bottle?

2. The following table represents the cost per foot of ribbon, but some of the numbers have rubbed off. Find the missing values:

Ribbon (Feet)	Cost (Dollars)
1	
3	$13.95
	$23.25
10	

3. Liam paints four rooms of an apartment complex in seven days. If he continues at that rate, how many rooms will he paint in 35 days?

4. Barry the stonemason paid $3,460 for 400 pounds of polished pebbles. Fill out the table below and use multiplication and division relationships to determine how much he would have paid for 100 pounds and how much he could have purchased for $5,000:

Price in dollars			
Quantity in pounds			

Using Ratio Tables for Unit Conversions

Unit conversions are another type of problem where ratio tables can be used to determine solutions. A unit conversion is when you want to change from using one unit of measurement to a different unit of measurement. You may want to convert feet into miles, ounces into pounds, or, as in the following example, from kilograms into pounds.

Example: *Beckett goes to the doctor, and the doctor says he weighs 53 kilograms. Help Beckett figure out how much he weighs in pounds*

Solution: We will use a conversion rate that changes pounds to kilograms in a ratio table. The conversion rate for changing kilograms to pounds is 2.2 kilograms for every pound. We'll begin this problem by putting the conversion rate in the first column of the table and Beckett's weight in kilograms in the second column of the table:

Kilograms	1 kilogram	45 kilograms
Pounds	2.2 pounds	? pounds

Next, figure out the multiplication relationship that exists between the columns in the rate table. Since 1 kilograms × 45 = 45 kilograms, the multiplication relationship is a multiple of 45. Multiply 2.2 pounds by 45 in order to calculate the equivalent weight in pounds: 45 × 2.2 = 99 pounds:

Kilograms	1 kilogram	45 kilograms
Pounds	2.2 pounds	99 pounds

Using Division to Find the Multiplication Relationship

Finding the multiplication relationship might not always be so easy. Consider the following instance, where a nurse tells you that you are 145 centimeters tall and you want to know your height in inches. You would first have to know the conversion rate that every inch is 2.5 centimeters. Organize your conversion ratio and given information in a ratio table:

Centimeters	2.5 centimeters	145 centimeters
Inches	1 inch	? inches

Now you are looking for the multiplication relationship between 2.5 and 145, which is not as obvious to calculate as in our previous example. In order to determine the multiplication relationship, divide 145 by 2.5: 145 ÷ 2.5 = 58. Now multiply one inch by 58 to determine that your height is 58 inches (4 feet, 10 inches) tall.

When working with unit conversion tables, some unit conversions will have an obvious multiplication relationship, while other unit conversions will require you to divide in order to determine the correct multiplication relationship.

Practice 2

1. A gallon of water weighs about 8.3 pounds. Use the ratio table to determine the weight of a 500-gallon water tank.

Gallon	1 gallon	500 gallons
Pounds	8.3 pounds	? pounds

2. A soup recipe calls for 40 ounces of potatoes. If there are 16 ounces in a pound, use the ratio table to determine how many pounds of potatoes are needed for this recipe.

Ounces	16 ounces	40 ounces
Pounds	1 pound	? pound

3. The summit of Mt. Everest is at an altitude of 29,029 feet. If there are 5,280 feet in a mile, how many miles above sea level is the summit of Mt. Everest?

4. Maya bought $1\frac{3}{4}$ pounds of organic papayas at the Mar Vista farmers market for $7. She uses 14 ounces of it that same afternoon and then gives the remaining papayas to her neighbor. What is the value of the papayas Maya's neighbor received? (Recall that 1 pound equals 16 ounces.)

Rate Tables on Coordinate Planes

Graphs are one of the most common ways to represent rates and relationships between data points. They are particularly useful for making comparisons. Since graphs are used and misused all the time in the media, being able to read and interpret them is an essential life skill. We will plot information from a rate table on a coordinate plane.

Coordinate Plane Basics

Points plotted on the coordinate plane are written as (x,y) coordinate pairs. The top row of numbers in a horizontal table contains the x-coordinates and you will find their corresponding y-coordinates in the bottom row. When plotting points on a coordinate plane, keep the following information in mind:

1. A coordinate plane has two *axes*, which serve as two crossed number lines.
2. The x-axis is a horizontal number line with zero in the middle, positive numbers to the right, and negative numbers to the left.

3. The y-axis is a vertical number line with zero in the middle, positive numbers extending upward and negative numbers extending downward.
4. The point of intersection between the x and y axes is called the **origin**. It has an x value of 0 and a y value of 0. It is written (0,0):

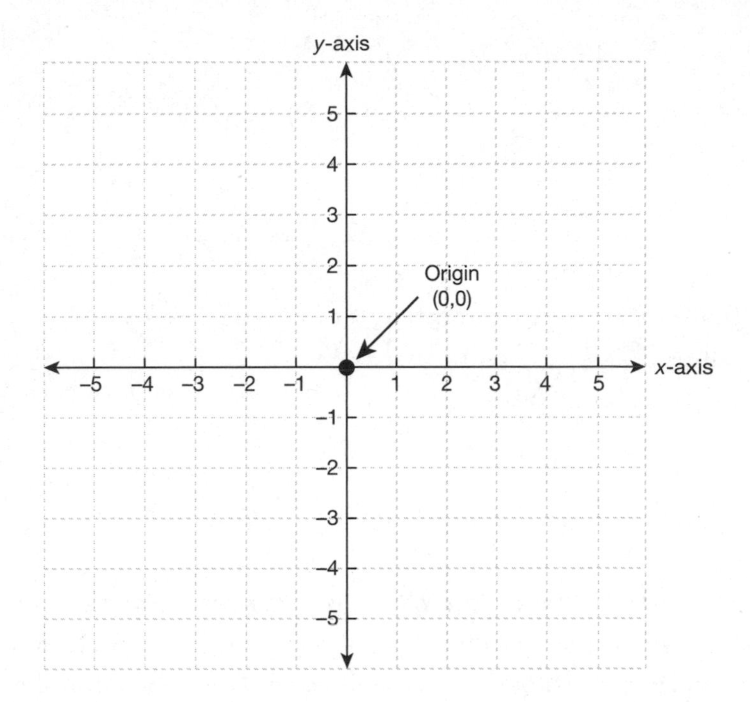

5. The x-coordinate is always the first term in the parentheses, (x,y). It gives instructions for how many units to move left or right. Starting at the origin, move *right* x units for positive values of x or move *left* x units for negative values of x.

6. The y-coordinate is always the second term in the parentheses, (x,y). It gives instructions for how many units to move up or down. After moving left or right for your x-coordinate, move y units up for a positive value of y, or move y units down for a negative value of y.

Look at the following graph and observe that point A is (4,–1) since its x-coordinate is 4, and its y-coordinate is –1. Point B is (–3,0) and it sits on the x-axis.

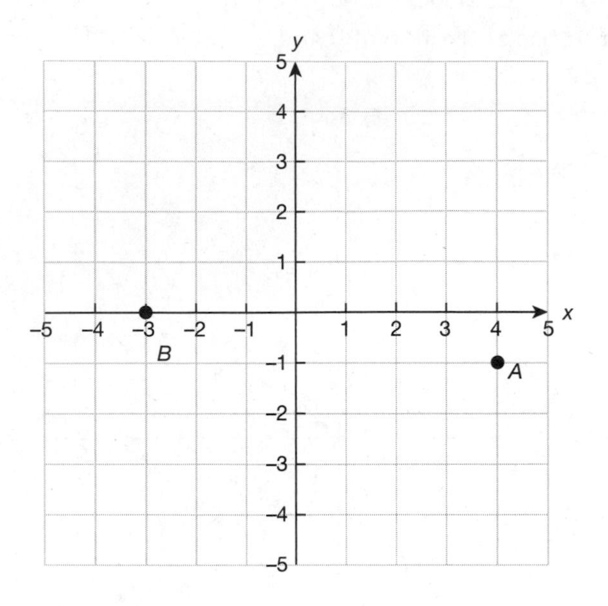

Plotting Rate Tables on the Coordinate Plane

Suppose the following rate table is used to show how much Tammy spent at the ice cream truck on ice cream sandwiches for the neighborhood kids over the course of a week:

Number of Kids	Money (Dollars)
2	$6
5	$15
3	$9

The information in the chart can be graphed on a coordinate plane after it is translated into coordinate pairs: (2,6), (5,15), and (3,9):

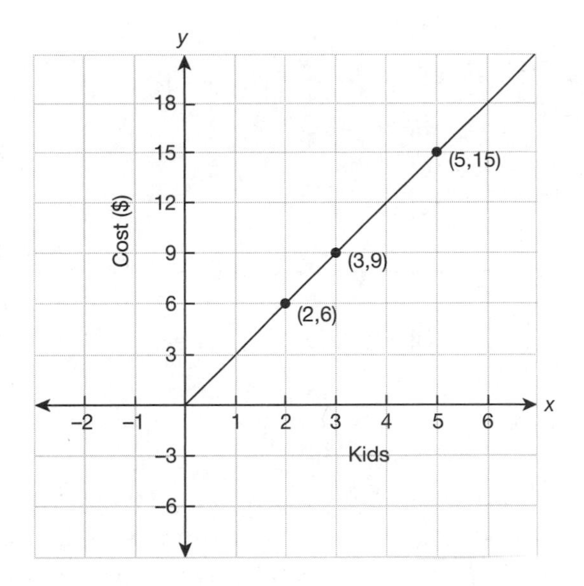

Notice that not much of the negative portions of the x- and y-axis are shown on this graph. Since the negative values are not relevant to the information in the table, it's appropriate not to include them in the graph. Also notice how the y-axis counts by three in order to accommodate the larger y-values. As long as you count by the same number throughout an axis, you can use whatever scale best fits the data.

Pulling Information Out of Graphs

Graphs are useful tools for finding out information. If you wanted to know how much Tammy would have spent on four kids' ice cream on a given day, you could find this out by going to the 4 on the x-axis and then moving up vertically until you hit the line. Then move horizontally until reaching the y-coordinate that corresponds with four kids. You can see in the following graph that Tammy would have spent $12.

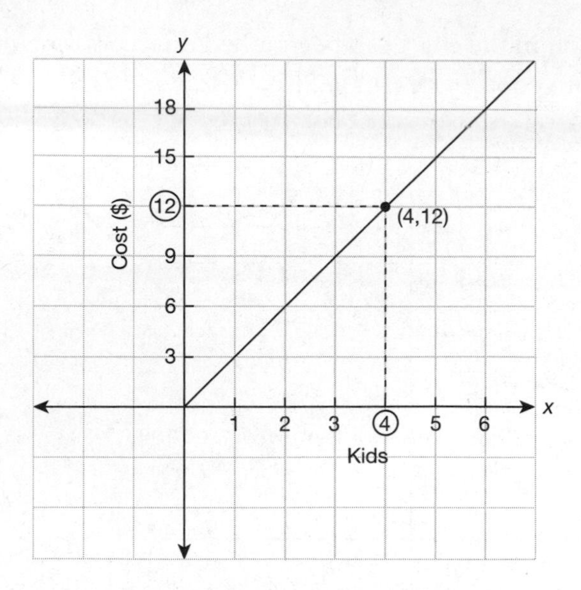

You could similarly see how many kids Tammy could treat if she had \$18 to spend. Find \$18 on the *y*-axis, move horizontally until meeting the line, and then move vertically down to arrive at six kids on the *y*-axis. This process is illustrated here:

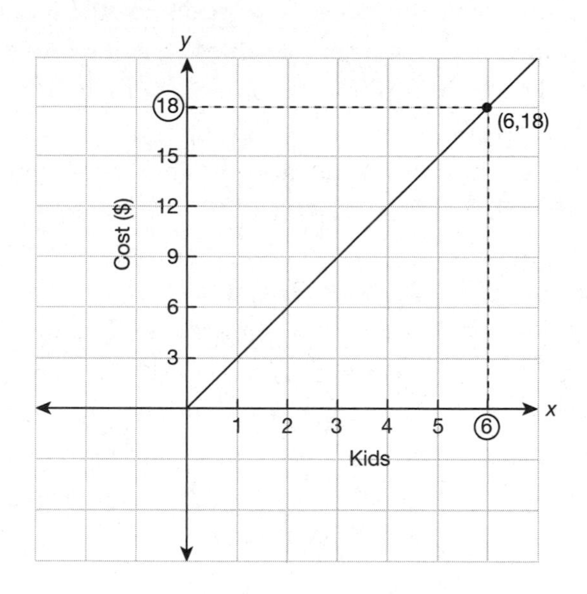

Practice 3

1. Graph the first two values in the table on a coordinate plane and make a line. Then use the line to find the missing values in the table.

Hot Cocoa	Cost
1	$2.50
4	$10
	$15
11	

2. What was the unit cost for a cup of hot cocoa? Where is this point shown on the graph?

The following graph and table model the number of dozens of muffins baked at Jamaica Cakes. Use these for questions 3 through 5.

Number of Hours	Number of Dozen
2	7
	14
5	17.5
6	

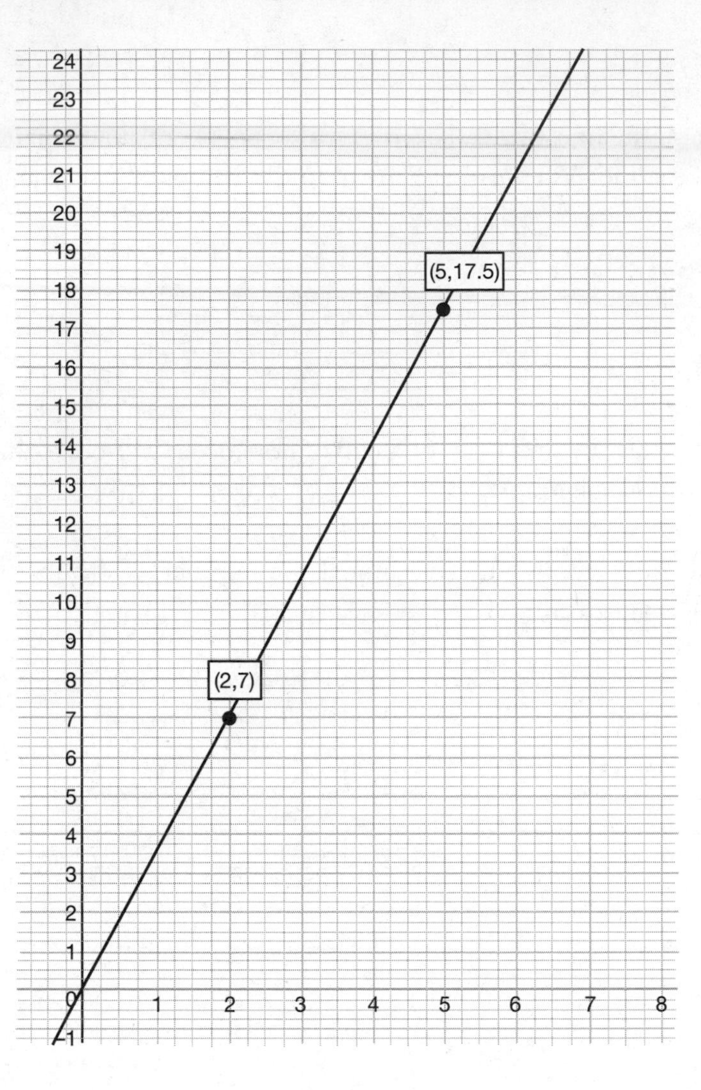

3. What should the x-axis and y-axis be labeled? Would it ever make sense to plot these on the opposite axes? Explain your reasoning.

4. What is the unit rate at Jamaica Cakes for baking muffins? How many dozen muffins do they bake every hour? Can you find this as a point on the graph?

5. Use the graph to fill in estimates for the missing values in the table.

Answers

Practice 1

1. Here we can use a division relationship instead of a multiplication relationship. We do this because the starting ratio has larger terms than the desired ratio. $138 ÷ $57.50 = 2.4. Now use that same division relationship to find the ounces: 3.4 ÷ 2.4 = 1.4 ounces.

2.

Ribbon (Feet)	Cost (Dollars)
1	$4.65
3	$13.95
5	$23.25
10	$46.50

3. $\frac{4 \text{ rooms}}{7 \text{ days}} = \frac{20 \text{ rooms}}{35 \text{ days}}$. So Liam will paint 20 rooms in 35 days.

4.

Price in dollars	$3,460	$865	$5,000
Quantity in pounds	400	100	578

Practice 2

1.

Gallon	1 gallon × 500 =	500 gallons
Pounds	8.3 pounds × 500 =	4,150 pounds

2.

Ounces	16 ounces × 2.5 =	40 ounces
Pounds	1 pound × 2.5 =	2.5 pounds

3.

Miles	1 mile × 5.5 =	5.5 miles
Feet	5,280 feet × 5.5 =	29,040 feet

4. First use a ratio table to figure out how many ounces of papayas Maya bought at the market:

Pounds	1 pound × 1.75 =	1.75 pounds
Ounces	16 ounces × 1.75 =	28 ounces

Maya bought 28 ounces of papayas and used 14 ounces, leaving 14 ounces to give to her neighbor. Therefore, Maya's neighbor received half the value of her $7 of papaya, or $3.50 worth of papayas.

Practice 3

1.

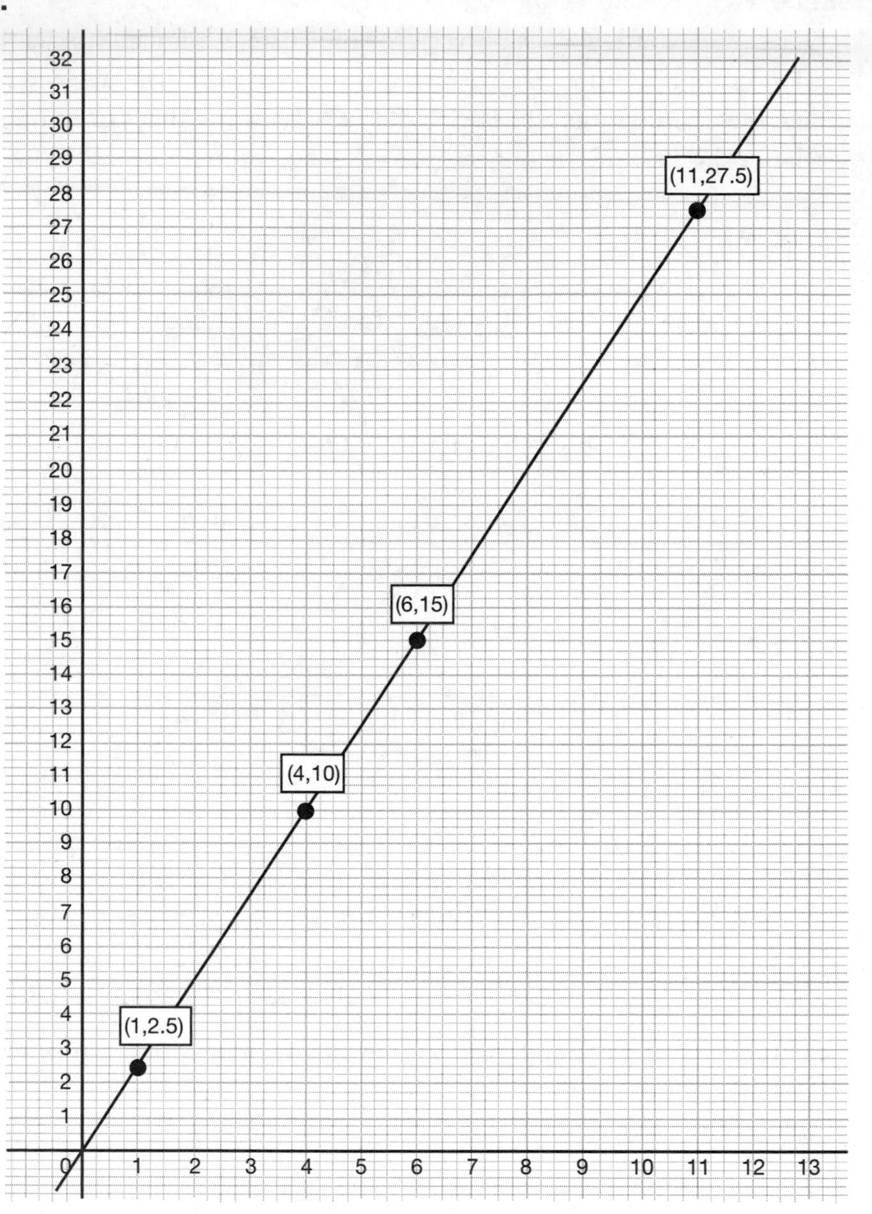

Hot Cocoa	Cost
1	$2.50
4	$10
6	$15
11	$27.50

2. A cup of hot cocoa costs $2.50 and this point is at (1,2.5).
3. The x-axis is hours and y-axis is number dozen. The number dozen could be on the x-axis if a bakery has to bake for as long as necessary to fulfill orders each day. Then, the number dozen would be determining time.
4. The unit rate is 3.5 dozen per hour and this point is illustrated by (1,3.5)

5.

Number of Hours	Number of Dozen
2	7
4	14
5	17.5
6	21

3

Using Rate to Work with Percents

STANDARD PREVIEW

In this lesson we will cover **Standard 6.RP.A.3.C**. You will learn how to use rate reasoning and ratio tables to solve different types of real-world problems involving percents.

Percentages: A Special and Very Common Use of Proportions

The word *percent* can be easily broken down into its two parts to understand its meaning: *per* means "for every" and *cent* means "100." Therefore, *percent* means "for every 100" and it is a special ratio with a denominator of 100. You should remember that ratios can be *part-to-part* or *part-to-whole*. Percentage is a part-to-whole ratio: the number before the percent symbol (%) is the part and the whole is 100. 20% means 20 out of 100. So, if 20% of a shipment of lightbulbs arrived broken, that means that 20 out of every 100 lightbulbs were broken.

There are three different basic types of percentage questions we will focus on in this lesson:

1. Using a ratio to determine the percentage that ratio represents *(There are 10 boys and 15 girls in a class. What percentage of the class is boys?)*
2. Using percentage to find a part when the whole is given. *(60% of voters in a town reported they will vote for Candidate A in the upcoming election. If the town has 2,000 people, how many votes will Candidate A receive?)*
3. Using percentage to find a whole when the part is given. *(20% of flights out of Sacramento were delayed due to weather. If 14 flights were delayed, what was the total number of flights?)*

Using a Ratio to Find a Percentage

Since a percentage is a ratio out of 100, any ratio can be converted into a percentage. Let's return to the example we began with in Lesson 1. A class of 25 students has 10 boys and 15 girls. Let's figure out what percentage of the class is boys by setting up a ratio table where the *part* is represented by *boys* and the *whole* is represented by *students*.

Part	10 boys	?
Whole	25 students	100

The multiplication relationship between the given ratio and the new percentage ratio will be multiplication by four:

Part (Boys): $10 \times 4 = \mathbf{40}$
Whole (Students): $25 \times 4 = \mathbf{100}$

Since the ratio $\frac{10}{25}$ is equivalent to $\frac{40}{100}$, it is determined that boys make up 40% of this class.

Making Percentages Out of Friendly Ratios

When the *whole* of a ratio is a number that is a factor of 100, like four, 10, 20, 25, or 50, it is much easier to recognize the multiplication relationship between the given ratio and a percentage ratio out of 100. We call these types of ratios *friendly ratios* since they don't make us work so hard to convert them into percents. Here are a few examples of friendly ratios:

- A ratio of 4 to 5 is equivalent to a ratio of 80 to 100 (80%) by multiplying 4 and 5 by 20:

  $\frac{4}{5}\left(\frac{\times 20}{\times 20}\right) = \frac{80}{100} = 80\%$

- A ratio of 9 to 10 is equivalent to a ratio of 90 to 100 (90%) by multiplying 9 and 10 by 10:

  $\frac{9}{10}\left(\frac{\times 10}{\times 10}\right) = \frac{90}{100} = 90\%$

- A ratio of 31 to 50 is equivalent to a ratio of 62 to 100 (62%) by multiplying 31 and 50 by 2:

  $\frac{31}{50}\left(\frac{\times 2}{\times 2}\right) = \frac{62}{100} = 62\%$

In the previous examples, each ratio was represented as a fraction and then multiplied by another fraction that was equivalent to 1 in order to create an equivalent percentage fraction. $\frac{20}{20} = \frac{10}{10} = \frac{2}{2} = 1$, since they have the same number in the numerator and denominator. Remember that multiplying any number by one will never change the *value* of that number, but will instead create an *equivalent* fraction. (You probably recognize this method from creating equivalent fractions when adding and subtracting fractions.)

Reducing Relationships into Friendly Ratios

But what can we do about ratios like $\frac{14}{35}$ that don't have a factor of 100 in the denominator? Although this ratio does not look so friendly, in both life and math, looks can be deceiving! Sometimes we can reduce a ratio with an undesirable denominator to uncover a friendly ratio. For example, if 14 out of 35 students were late to school, $\frac{14}{35}$ can be reduced to $\frac{2}{5}$ before turning it into a percentage. Once $\frac{14}{35}$ is rewritten as $\frac{2}{5}$, we can simply use a multiplication relationship of 20 to get our equivalent percentage ratio:

$\frac{2}{5}\left(\frac{\times 20}{\times 20}\right) = \frac{40}{100} = 40\%$

Let's consider a class with 26 girls and 14 boys and find the percentage of girls in this class. Start by realizing that there are 40 students in total and put your given information in a ratio table:

Part	26 girls	?
Whole	40 students	100

Notice that the ratio $\frac{26}{40}$ can be reduced by dividing by two, which will create a friendly ratio:

Part	26 girls ÷ 2 =	13 boys	?
Whole	40 students ÷ 2 =	20 students	100

Now this friendly ratio can use a multiplication factor of 5 to create a percentage ratio out of 100:

Part	26 girls	13 boys × 5 =	65
Whole	40 students	20 students × 5 =	100

Since the ratio $\frac{26}{40} = \frac{13}{20} = \frac{65}{100}$, we can determine that 65% of the class members are girls.

Practice 1

1. Fill out the following table by turning the given ratios into percentages. The first row has been completed as an example:

Ratio	Fraction (Reduced)	Multiplication Factor	Percentage Ratio	Percentage %
12 out of 15	$\frac{12}{15}\left(\frac{\div 3}{\div 3}\right) = \frac{4}{5}$	$\times \frac{20}{20}$	$\frac{4}{5} \times \frac{20}{20} = \frac{80}{100}$	80%
1.5 out of 10				
3 out of 4				
18 out of 30				
60 out of 75				

2. Ella got 54 out of 60 questions correct on her math quiz. Find out her score as a percentage by using the following ratio table:

Part	54 correct	?
Whole	60 total	100

Use the following information to complete questions 3 through 5:

A statistic reports that last year the average American used 167 disposable water bottles, but only recycled 38.

3. Fill out the ratio table and determine what the multiplication relationship is between the given statistic and the percentage ratio. What do you notice about it?

Part	38 recycled	
Whole	167 used	100

4. Use the ratio table above to determine what percentage of disposable water bottles were recycled by the average American according to the statistic.

5. Use your answer from question 4 to determine the percentage of disposable water bottles the average American threw away. (Hint: Use the fact that your percentage from question 4 represents that number of bottles out of 100.)

6. In Portland, Oregon 17,000 workers returned a survey about biking to work. 1,020 of them reported that they bike to work on a regular basis, regardless of the weather. Make and solve a ratio table to determine what percentage of workers are bike commuters.

Using a Percentage to Find a Part

Newspapers often poll groups of people and report the results in percentages. In one poll, a newspaper asked 2,000 people their preferences in an election. 60% of respondents said they are planning on voting for Candidate A. You can figure out the number of votes 60% means by using the number of people who participated in the survey. We'll use a ratio table with 60% input as 60 out of 100:

Part	60	?
Whole	100	2,000

First, find the multiplication relationship between the ratios. Since $100 \times 20 = 2{,}000$, we can multiply 60×20 to find out how many people said they are planning to vote for Candidate A: $60 \times 20 = 1{,}200$.

Let's consider another town voting in the same election where the multiplication relationship is not as easy to recognize. Suppose that in a smaller town of 1,640 people, 75% of the voters have said they will vote for Candidate A. Will Candidate A get more or less votes in this smaller town? Remember that 75% means $\frac{75}{100}$, so begin by filling that ratio into your ratio table:

Part	75	?
Whole	100	1,640

First, find the multiplication relationship between the ratios. Since it isn't clear what number should be multiplied by 100 to get 1,640, divide 1,640 by 100 to find the multiplication relationship: 1,640 ÷ 100 = 16.4. Now multiply both parts of the ratio by 16.4 to create the equivalent percentage ratio:

$$100 \times 16.4 = 1{,}640$$
$$75 \times 16.4 = 1{,}230$$

So 1,230 people have said they will vote for Candidate A. Notice that even though the second town of 1,640 is smaller than the first town of 2,000, Candidate A will actually get more votes there because of the percentage of people who support his policies.

What can we do if the multiplication relationship going from one number to a second number is not easy to recognize? Dividing the second number by the first number will reveal the multiplication relationship.

Using a Percentage to Find the Whole

You can also use percentages to find the whole when you know a part. For example, suppose 20% of flights out of Sacramento were delayed due to weather. If we know that 14 flights were delayed, how could we determine the total number of flights scheduled out of Sacramento that day? Let's set up a ratio table to model this information and find the multiplication relationship. Notice that 14 has been filled in as the *part*:

Part	20	14
Whole	100	?

Divide 14 by 20 to find the multiplication relationship: $14 \div 20 = 0.7$. Now use 0.7 to create an equivalent ratio:

Part	$20 \times 0.7 =$	14
Whole	$100 \times 0.7 =$	70

Since $100 \times 0.7 = 70$, there were 70 flights out of Sacramento.

Practice 2

1. Sometimes it is easy to recognize the multiplication relationship in a ratio table. What do you do to find the multiplication relationship when it is not easy to recognize?

2. Little Mikey eats everything! He ate 25% of the donuts at this morning's meeting! If he ate five donuts, how many donuts were there to begin with at the meeting?

3. It is estimated that 2% of the world's population has green eyes. Roughly how many people will have green eyes when the population reaches 8 billion people?

4. If approximately 8% of people have blue eyes, use *your answer* from question 3 to estimate the number of people who will have blue eyes when the population reaches eight billion. *Do not use a ratio table.* (*Hint:* What is the multiplication relationship between 2% and 8%?)

5. Samantha bought a stock that crashed the month after she purchased it. It is now worth $63 per share, which is just 30% of what she originally paid for it. Use a ratio table to determine the original price of the stock.

Answers

Practice 1

1.

Ratio	Fraction (Reduced)	Multiplication Factor	Percentage Ratio	Percentage %
12 out of 15	$\frac{12}{15}\left(\frac{\div 3}{\div 3}\right) = \frac{4}{5}$	$\times \frac{20}{20}$	$\frac{4}{5} \times \frac{20}{20} = \frac{80}{100}$	80%
1.5 out of 10	$\frac{1.5}{10}$	$\times \frac{10}{10}$	$\frac{1.5}{10} \times \frac{10}{10} = \frac{15}{100}$	15%
3 out of 4	$\frac{3}{4}$	$\times \frac{25}{25}$	$\frac{3}{4} \times \frac{25}{25} = \frac{75}{100}$	75%
18 out of 30	$\frac{18}{30}\left(\frac{\div 6}{\div 6}\right) = \frac{3}{5}$	$\times \frac{20}{20}$	$\frac{3}{5} \times \frac{20}{20} = \frac{60}{100}$	60%
66 out of 75	$\frac{66}{75}\left(\frac{\div 3}{\div 3}\right) = \frac{22}{25}$	$\times \frac{4}{4}$	$\frac{22}{25} \times \frac{4}{4} = \frac{88}{100}$	88%

2. Divide 54 and 60 by six to reduce it to the friendly ratio $\frac{9}{10}$. Multiplying $\frac{9}{10}$ by $\frac{10}{10}$ gives a ratio of $\frac{90}{100}$, so Ella got 90% of the questions correct. Nice going, Ella!

Part	$54 \div 6 = 9$	90
Whole	$60 \div 6 = 10$	100

3. The multiplication relationship is determined by $100 \div 167$ so it is less than one. The multiplication relationship is approximately 0.60.

4. $\frac{22.8}{100}$ or 22.8% of water bottles were recycled by the average American.

Part	$38 \times 0.60 =$	22.8%
Whole	$167 \times 0.60 =$	100

5. Since 22.8% of disposable water bottles were recycled, subtract that from 100% to determine that the average American threw away 77.2% of the disposable water bottles they used. Time to get reusable water bottles, America!

6. 6% of workers who filled out the survey are bike commuters:

Part	1,020 bikers $\times \frac{100}{17{,}000} =$	6
Whole	17,000 workers $\times \frac{100}{17{,}000} =$	100

Practice 2

1. If the multiplication relationship going from one number to a second number is not easy to recognize, dividing the second number by the first number will reveal the multiplication relationship.
2. Divide 20 by 100 to uncover the multiplication relationship of 0.2. There were 20 donuts to begin with.

Part	$25 \times 0.2 =$	5
Whole	$100 \times 0.2 =$	20

3. Approximately 160 million people will have green eyes.

Part	$2 \times 80{,}000{,}000 =$	160,000,000
Whole	$100 \times 80{,}000{,}000 =$	8,000,000,000

4. Since 8% is four times bigger than 2%, multiply 160 million by four to estimate that 640,000,000 people will have blue eyes when the population reaches eight billion.
5. Samantha originally purchased the stock for $210. Yikes!

Part	$30 \times 2.1 =$	$63
Whole	$100 \times 2.1 =$	$210

4

Exponents and Order of Operations

STANDARD PREVIEW

In this lesson we will cover **Standard 6.EE.A.1**. After a brief review of operations with signed numbers, you will learn how to write and evaluate numerical expressions with exponents. Then you will learn the correct order in which to perform arithmetic operations such as addition, subtraction, multiplication, and division.

Review of Operations with Signed Numbers

Before we get started with exponents and order of operations we'll do a quick review of arithmetic operations with signed numbers. It's important that you feel comfortable adding, subtracting, multiplying, and dividing positive and negative numbers. Why? Because negative numbers come up in life just as much as positive numbers. Stock prices fall, money gets spent, and temperatures drop—all of these are illustrations of negative numbers in action. Since this is an algebra book and we've decided not to focus on the standards that fall into The Number System category, this review will be brief.

Adding Like Signs

Think of positive numbers as money you've earned and negative numbers as money you've spent. It shouldn't be a surprise that if you earn \$20 and then you earn \$10, you have earned \$30. A positive plus a positive always equals a bigger positive. Conversely, you can probably understand that if you spend \$10 and then you spend \$5, you just spent \$15. This scenario is represented as (–\$10) + (–\$5) = –\$15. (The parentheses in this equation do not have any mathematical significance—they simply compartmentalize each negative number. Since it is important to feel comfortable with the various ways math is written, we will sometimes use the extra parentheses and other times omit them throughout this book.) The scenario of spending money twice can be generalized as the rule that a negative plus a negative equals a bigger negative.

Positive + Positive = Positive; \$20 + \$10 = \$30

Negative + Negative = Negative; (-\$10) + (-\$5) = -\$15

Adding Negatives and Positives

When you earn money and then you spend money, do you end up with money left over or do you end up in debt? This *depends* on how much money you earned and how much money you spent. If you earned more money than you spent, you will have money left over (which is a positive number). But, if you spend more money than you earned, you will end up with a debt, which is a negative number. For example, \$20 + (–\$25) models a situation where you earned \$20 and then spent \$25. It's probably clear to you that in this case, you would be \$5 in debt, or have –\$5. Therefore, \$20 + (–\$25) = –\$5. *To add a negative and positive number: ignore the signs, subtract the numbers, and give the difference the sign of the larger number.* So, to perform 20 + (–25), we subtracted 20 from 25 and gave the difference of 5 the negative sign from the larger number, –25. This equation has the same answer if the order of the numbers is switched around: –25 + 20 = –5. In addition problems, it is not significant if the negative number is on the left or right.

Negative + Positive = Depends! Subtract the two numbers and give your answer the sign of the larger number.

Subtracting Signed Numbers

Don't do it! It is way too easy to make careless errors when subtracting mixed numbers. Life is tough enough, and there are many things out there that will try to ruin your day, so don't allow subtracting signed integers be one of them! Look at how confusing the expressions 5 – (–3) and –26 – (–8) are. Hopefully you know that *subtracting a negative is the same as adding a positive. In fact, subtraction can always be turned into addition by adding the opposite*. For example 10 – 9 is the same thing as 10 + (–9). We will refer to this as the *keep-switch-switch* technique: *keep* the sign of the first number, *switch* the subtraction to addition, and *switch* the sign of the second number. If you use *keep-switch-switch* to turn subtraction into addition, you just need to remember your rules for adding signed numbers and you don't have to worry about another set of rules for subtracting signed numbers. Of course we don't want you to start turning easy things like 10 – 2 into 10 + –2, but *keep-switch-switch* will help turn something odd looking like 5 – (–3) into 5 + (+3). Similarly, the expression –26 – (–8) becomes –26 + 8 after using *keep-switch-switch*. Once you have rewritten your subtraction problem as addition, follow the rules for addition.

Turn odd-looking subtraction problems into addition by using *keep-switch-switch*:

- *keep* the sign of the first number
- *switch* the subtraction to addition
- *switch* the sign of the second number

Then follow the rules for addition.

Multiplying and Dividing Signed Numbers

Multiplication and division follow the same rules for signed numbers. When you multiply or divide opposite signs, the answer will be negative. Think of 4 × –\$5 as being a way to illustrate that you *spent \$5* on 4 separate occasions. In total you would have spent \$20, which is expressed as –20. Therefore, 4 × –5 = –20. When you multiply or divide the same signs, the answer will be positive. Think of $-4 \times -\$5$ as representing someone forgiving a \$5 debt you owe them four times in a row. (That would be like you earning \$20.) Therefore, $-4 \times -\$5 = 20$ and the same relationship holds true for division: $-8 \div -4 = 2$.

Negative ×/÷ Positive = Negative; $4 \times -5 = -20$

Negative ×/÷ Negative = Positive; $-4 \times -\$5 = 20$

What Are Exponents?

Now that you have brushed up on your signed numbers, let's move on to exponents. You will be able to easily recognize when an **exponent** is being used because an exponent is written as a small number above the upper right-hand corner of a regular-sized number, which is called the **base**. In the expression 3^4, four is the exponent and three is the base. Exponents are a mathematical notation for representing repeated multiplication in shorthand. Instead of writing $3 \times 3 \times 3 \times 3$, an exponent of four can be used to indicate that the number three should be multiplied by itself four times:

$$3 \times 3 \times 3 \times 3 = 3^4$$

The exponent always gives instructions for how many times the base will be multiplied by itself. Therefore, 10^6 means 10 multiplied by itself 6 times which is 1,000,000. A common mistake is to think that 10^6 means 10×6, but that would give you 60, a very different answer from 1,000,000!

An exponent is used to indicate how many times the base should be multiplied by itself.

$$2^4 = 2 \times 2 \times 2 \times 2 = 16$$

ERROR ALERT! Many students *know* that an exponent means to multiply the base by itself the number of times dictated by the exponent, but it is all too easy to make careless errors with them. $24 \neq 8$! Do not multiply the base times the exponent!

The Language of Exponents

Exponents are normally referred to as **powers**. The expression 3^4 is said *three to the fourth power* or *three raised to the power of four*. The exponents two and three have their own special names:

Squared: When a base is raised to a power of two, the base is being *squared*. 5^2 is said *five squared*. Since $5^2 = 5 \times 5$, write $5^2 = 25$.
Cubed: When a base is raised to a power of three, the base is being *cubed*. 5^3 is said *five cubed*. Since $5^3 = 5 \times 5 \times 5$, write $5^3 = 125$.

Example: *Write $4 \times 4 \times 4$ in exponential form and evaluate it.*
Solution: Since 4 is being multiplied by itself four times, $4 \times 4 \times 4 = 4^3 = 64$

Example: *Write "two to the power of five" in exponential form and evaluate it.*
Solution: *2 to the power of five* is 2^5: $2 \times 2 \times 2 \times 2 \times 2 = 32$

Practice 1

1. Write *10 cubed* in exponential form and evaluate it.

2. Find two different exponential expressions that have a value of eight.

3. What is the value of *negative three squared*?

4. What is the value of *negative three cubed*?

5. Compare and contrast your answers to questions 3 and 4.

6. What is the value of *five squared*?

7. What is the value of *negative five squared*?

8. Compare and contrast your answers to questions 6 and 7.

Order of Operations: PEMDAS

In order to correctly solve problems that include more than one operation, there is a specific order of operations that must be followed. These rules ensure that everyone in the world is doing the same steps in the same order when solving a given sequence of operations. It would be really weird if students from one country determined that $1 + 2 \times 4 = 12$ and students from a different country determined that $1 + 2 \times 4 = 9$. (Which of those do you think is correct?)

The correct Order of Operations to follow when evaluating expressions is:

1. **P**arentheses
2. **E**xponents and Roots*
3. **M**ultiplication and **D**ivision*
4. **A**ddition and **S**ubtraction*

*These operations are performed as they arise from left to right in a problem.

This sequence is commonly referred to by the acronym *PEMDAS*. You can remember *PEMDAS* by reciting *Please Excuse My Dear Aunt Sally* or *Purple Elephants Marching Down A Street* or *Pink Elephants Make Delicious Afternoon Snacks*. However you remember it . . . remember it! It will help you correctly evaluate mathematical expressions, but only if you pay attention to the small, but *very* important asterisk at the end of the list above! It is critical that you treat the operations that are on the same line together as equal partners that are evaluated from left to right. Doing multiplication before division can get you in some hot water—read on!

ERROR ALERT! The most common mistake that students make is always doing multiplication before division. Please do NOT make this mistake! Notice that in solving $24 \div 6 \times 2$, you will get two different answers depending on whether you (incorrectly) do multiplication *before* division, or do multiplication and division from *left to right*:

- Multiplication *before* division: $24 \div 6 \times 2 = 24 \div 12 = 2$ (Incorrect!)
- Multiplication and division from *left to right*: $24 \div 6 \times 2 = 4 \times 2 = 8$ (Correct!)
- Notice that the two different methods give two different answers, so only one method can be correct: moving from left to right when performing multiplication and division. Beware!

PEMDAS in Action

Now that you have *read* about the order of operations, let's walk through some examples together to see them in action. It is a good idea to write each example down on paper and try to work it out on your own before looking through the following detailed steps. This way you will be able to identify your errors, and pay close attention that you avoid these errors in the future.

Example: *Evaluate* $4(2)^2 + 3(1 - (-2))$

Solution: While it may be tempting to go from left to right, we cannot multiply 2 by 4 and then square it because Exponents come before Multiplication in the order of operations. Plus, we also have Parentheses in the second half of the problem, and the order of operations indicates we do that FIRST. Always. Let's go through the steps one by one:

1. Parentheses: $(1 - (-2))$. Subtracting a negative number is like adding a positive, so turn the subtraction into addition by using *keep-switch-switch*: $(1 - (-2))$ becomes $(1 + 2)$ which equals 3. The problem now reads: $4(2)^2 + 3(3)$.
2. Exponents: $(2)^2$. This is 4. The problem now reads: $4(4) + 3(3)$.
3. Multiplication: $4(4) = 16$ and $3(3) = 9$. The problem now reads: $16 + 9$.
4. Addition: $16 + 9$, which equals 25. Therefore, the answer is 25.

Example: *Evaluate* $\frac{12 + 8}{12 - 4^2}$

Solution: While it may be tempting to cancel out the 12 in the numerator with the 12 in the denominator, the division bar acts as a set of parentheses. Therefore, this equation must be performed as $\frac{(12 + 8)}{(12 - 4^2)}$. The steps are:

1. **P**arentheses: (12 + 8) = 20 in the numerator. To do the denominator we must evaluate the Exponent before Subtraction. The problem now reads: $\frac{20}{12-4^2}$
2. **E**xponents: $4^2 = 16$. The problem now reads: $\frac{20}{12-16}$
3. **S**ubtraction: 12 – 16 = –4. The problem now reads: $\frac{20}{-4}$
4. **D**ivision: 20 ÷ –4 = –5, so –5 is the final answer.

Practice 2

Simplify the following expressions.

1. $3^2 - 2^3$
2. Compare the answer to the expression $(5 - 3) \times (4 + 4 \div 2)$ to the answer to the expression $5 - 3 \times 4 + 4 \div 2$.
3. $\frac{8^2 + 22 \div 2}{12 - 3^2}$
4. $15 - \frac{1}{2}(4 + 2)^2$
5. $\frac{5^4}{5^3} + \frac{2^5}{2^4} \times \frac{4^3}{4^2} \div 2$
6. $1^9 \times \frac{2^8}{4^4}$
7. What is true about the value of one risen to any power?
8. Determine whether it is appropriate to express *eight tripled* as 8×3 or 8^3 and find the value of *eight tripled*.

Answers

Practice 1

1. $10^3 = 10 \times 10 \times 10 = 1{,}000$
2. $2^3 = 8$ and $8^1 = 8$. The exponential expressions 2^4 and 4^2 do *not* have a value of eight.
3. *Negative three squared* is $(-3)^2 = -3 \times -3 = 9$
4. *Negative three cubed* is $(-3)^3 = -3 \times -3 \times -3 = -27$
5. When –3 is squared the answer is positive, but when –3 is cubed the answer is negative.
6. *Five squared* is $5^2 = 5 \times 5 = 25$
7. *Negative five squared* is $(-5)^2 = -5 \times -5 = 25$
8. When both five and negative five are squared the answer is the same because the negative sign cancels out when –5 is multiplied by itself.

Practice 2

1. $3^2 - 2^3 = 9 - 8 = 1$
2. $(5 - 3) \times (4 + 4 \div 2) = 2 \times 6 = 12$ and $5 - \underline{3 \times 4} + \underline{4 \div 2} = 5 - 12 + 2 = -5$, so the answers are very different.
3. $\frac{(8^2 + 22 \div 2)}{(12 - 3^2)} = \frac{(64 + 11)}{(12 - 9)} = \frac{75}{3} = 25$
4. $15 - \frac{1}{2}(4 + 2)^2 = 15 - \frac{1}{2}(36) = 15 - 18 = -3$
5. Since this a long equation with three quotients, it is a good idea to evaluate them individually and then plug them back into the equation:
   $\frac{5^4}{5^3} + \frac{2^5}{2^4} \times \frac{4^3}{4^2} \div 2$
   $\frac{5^4}{5^3} = 5$, $\frac{2^5}{2^4} = 2$, and $\frac{4^3}{4^2} = 4$, so plug these in:
   $5 + 2 \times 4 \div 2 = 5 + 8 \div 2 = 5 + 4 = 9$
6. $1^9 \times \frac{2^8}{4^4} = 1 \times \frac{256}{256} = 1$
7. One risen to any power will always be one since 1 times itself is 1.
8. The term *tripled* means multiplied by three, so *eight tripled* is $8 \times 3 = 24$.

5

Introduction to Algebraic Expressions

STANDARD PREVIEW

In this lesson we will cover Standards 6.EE.A.2.A and 6.EE.A.2.B, which contain some of the most important foundations of algebra. You will learn how to translate words into numerical expressions and you will also translate words into algebraic expressions with variables that stand for numbers

What *Is* Algebra?

In the last lesson we focused on arithmetic by practicing the Order Of Operations mapped out in *PEMDAS*. *Arithmetic* deals with operations on numbers while *algebra* uses variables, or letters, to create mathematical models to represent the world around us. These mathematical models are expressed as algebraic equations that can be solved to find an unknown value, or can be studied to identify trends and make predictions. If you have used the formula *Perimeter* = $4s$ to find the distance around a square, then you have already gotten your algebra feet wet.

English-to-Math Translations

Since we are going to be using algebra to model real-world situations, our first step is to learn some basics on how to translate words into math. Let's begin with four key words that represent addition, subtraction, division, and multiplication when translating word problems into mathematical sentences:

Sum: A *sum* of two or more numbers is the answer to an *addition* problem.

Example: The sum of three and five is eight (i.e., $3 + 5 = 8$).

Difference: A *difference* between two numbers is the answer to a *subtraction* problem.

Example: The difference of five and four is one (i.e., $5 - 4 = 1$).

Quotient: A *quotient* is the answer to a *division* problem.

Example: The quotient of 18 and six is three (i.e., $18 \div 6 = 3$).

Product: A *product* is the answer to a *multiplication* problem.

Example: The product of eight and nine is 72 (i.e., $8 \times 9 = 72$).

Don't expect to see the words above all the time—similar to how there are many ways to say "hello," there are lots of other ways to indicate the four operations. Think about $6 + 7$ and you'll realize that it could be *the sum of six and seven*, *six plus seven*, *six more than seven*, *six increased by seven*, or *the total of six and seven*. Each of these phrases contains a keyword that signals addition: *plus*, *more than*, and *total*. The four basic operations have a handful of keywords that act as clues to which operation will be used. The following chart lists several of the main keywords for each operation:

Addition	Subtraction	Multiplication	Division
sum	difference	product	quotient
combine	take away	times	percent
total	less than	of	out of
plus	minus	every	share
and	decrease	each	split
altogether	left	factors	average
increase	fewer	double (× 2)	each
more than	remove	triple (× 3)	per

Some Trickier Translations

In addition to these words that indicate the four basic arithmetic operations, keep your eyes out for certain words and phrases that combine more than one operation and may require parentheses:

The quantity of: *The quantity of* indicates that there are two or more terms combined to make one term. This combination of multiple terms into a single term requires *parentheses.*

Example: *Six times the quantity of five plus 10* is written $6(5 + 10)$.

The sum of and the difference of: These two terms are like *the quantity of* and are illustrated with parentheses.

Example: *Five times the difference of eight and three* is written $5(8 - 3)$.

Example: *The sum of 20 and 19 divided by three* is written $(20 + 19) \div 3$

Sometimes words like *from* and *less than* indicate subtraction, but the order of the terms must be reversed:

From: *Subtract eight from 10* is written as $10 - 8$ and *not* $8 - 10$. This is because the phrase indicates that 8 is the number being subtracted, not 10.
Less than: *Two less than 20* is written as $20 - 2$ and *not* $2 - 20$. The phrasing indicates that 20 is the starting number which is being reduced by 2.

These words represent multiplication by two or three:

Twice: *Twice as much as the original $50 price* is written as $2 \times \$50$
Triple: *Triple the amount of last year's 4,000 attendees* is written as $3 \times 4{,}000$

Practice 1

Represent each phrase as a numerical expression. Do not evaluate.

1. 10 passengers increased by 20 and then decreased by seven

2. $85 is shared by three siblings

3. Four less than the product of six and 12

4. Triple the sum of their $500 October sales and $700 November sales

5. They split the difference of her expenses of $3,200 and his expenses of $2,800

6. $40 less than $100 is increased by $80 and then doubled

Writing Phrases as Algebraic Expressions

Now that you have some practice translating words into numerical expressions, let's translate phrases into algebraic expressions. What is the difference between numerical and algebraic phrases? **Numerical phrases** contain only numbers while **algebraic phrases** contain at least one unknown quantity. In a verbal description, that unknown can often be referred to as *a number* such as in *five times a number*. If there is a second unknown it will usually be referred to as *another number* or *a second number*.

Variables

Although the variable x is most commonly used to represent the unknown quantity in algebraic expressions, any letter or symbol can be used. The phrase *five more than 10* is translated as $5 + 10$, and the phrase *five more than a number* could be written as $5 + x$, $5 + m$, or even $5 + K$. When a phrase refers to two unknown values, it is common to represent the first unknown number with x and the second unknown number as y. Using this convention, the phrase *twice a number plus four times another number* would be written $2x + 4y$.

Remember to look for the trickier works like *quantity*, *sum of*, and *difference of*, which indicate that two items should be grouped together with parentheses:

Example: *10 times the sum of five and a number*
The sum of indicates that $5 + x$ must be grouped together as a single term in parentheses: $10(5 + x)$

Practice 2

Represent each phrase as an algebraic expression.

1. The sum of five and twice a number w

2. One-third of the difference of six and a number

3. Thirty more than a number squared

4. The quantity 13 less than a number is tripled and then added to another number

5. Combine $48.90, $20.20, and a number, then cut it in half

6. A dozen fewer than a cubed number

Defining the Parts of Algebraic Expressions

Now that you are getting the feel for translating words into algebraic expressions, let's become familiar with some of the language used to discuss the building blocks of algebra. It is important that you know the meanings of these words since they will be used to explain more involved procedures with algebraic equations.

Constant: A fixed number that remains the same and does not change.

Example: In $y = 3x + 7$, 7 is the *constant*

Example: In $13x - \frac{1}{2}$, $\frac{1}{2}$ is the *constant*

Variable: A letter or symbol that represents a number in an algebraic expression.

Example: $-2x + 8y$; x and y are *variables*.

Coefficient: The number or symbol multiplied by a variable in an algebraic expression. It is very important to remember that when a variable is on its own, its coefficient is one:

Example: $4x^2 + 3x + 2$; 4 and 3 are coefficients, but 2 is a constant

Example: $x + 9y$; 1 is the coefficient of x and 9 is the coefficient of y

Term: A number, variable, or a coefficient multiplied by one or more variables. Terms in algebraic expressions are separated by addition or subtraction.

Example: $5t$; there is one term: $5t$

Example: $-12x + 3y - 10$; there are three terms: $-12x$, $3y$, and 10

Example: $-0.5x^2y - 10w$; there are two terms: $-0.5x^2y$ and $10w$

Factor: Numbers that are multiplied together in an algebraic expression—because of PEMDAS, terms that are added or subtracted within a set of parentheses are treated as a single factor.

Example: In $9xy$; there is one term with three factors: 9, x, y

Example: In $9(x + y)$; there are two factors: 9 and $(x + y)$

Algebraic expression: A mathematical sequence containing one or more variables or numbers connected by addition or subtraction.

Example: $-12x + 3y - 10$ is an algebraic expression with three terms

Example: $2L + 2W$ is an algebraic expression with two terms

You should be able to identify all the different parts of algebraic expressions, such as the constants, variables, coefficients, terms, and factors. Let's work through two examples together, but you might want to try to write down your answers first on separate paper before looking at the solutions so that you can see how well you understand these vocabulary words:

Example: Identify all the terms, factors, variables, coefficients, and constants in the algebraic expression $4w + x - yz + 7$.

Solution: $4w + x - yz + 7$ has four terms: $4w$, x, yz, and 7. The first term has two factors, 4 and w, and the third term has two factors, y and z. The variables are w, x, y, and z. The number 4 is a coefficient and 7 is a constant.

Example: How many terms and factors are in the algebraic expression $10 + 2r(4 + k)$?
Solution: Since there is only one addition sign, $10 + 2r(4 + k)$ has 2 terms: 10 and $2r(4 + k)$. (Notice how the parentheses means that we treat the $(4 + k)$ as one and it combines with $2r$ to be a single term.) $2r(4 + k)$ has 3 factors: 2, r, and $(4 + k)$.

Practice 3

1. Is 12 an algebraic expression? Is 12 a term? Is 12 a coefficient? How would you best define 12?

2. What is the difference between a constant and a coefficient? Give an example of each.

3. Explain the difference between a factor and a term. Then use $4xy$ as an example for your explanation.

4. What is the constant in $4x + 5y(6z - 7) - 2$?

5. In the algebraic expression, $12e + 8fg + 9h(i + 2j)$, how many *terms* are there? How many *factors* does each term have?

6. What is the sum of the coefficients in the algebraic expression $5x + y$?

Answers

Practice 1

1. $10 + 20 - 7$
2. $\$85 \div 3$
3. $(6 \times 12) - 4$
4. $3(\$500 + \$700)$
5. $\frac{1}{2}(\$3{,}200 - \$2{,}800)$ or $(\$3{,}200 - \$2{,}800) \div 2$
6. $(\$100 - \$40 + \$80)2$; same as $2(\$100 - \$40 + \$80)$

Practice 2

1. $5 + 2w$
2. $[\frac{1}{3}](6 - x)$ or $\frac{(6 - x)}{3}$
3. $30 + x^2$
4. $(x - 13) \times 3 + y$ or $3(x - 13) + y$
5. $\frac{(\$48.90 + \$20.20 + x)}{2}$
6. $x^3 - 12$

Practice 3

1. 12 is not an algebraic expression because it does not contain any variables. It is not a coefficient because it is not being multiplied to a variable. It is a term and it is also a constant.
2. A *constant* is a fixed number that is independent and not being multiplied to a variable. A *coefficient* is a number being multiplied to a variable. In the expression $13 + 6x$, 13 is a constant and 6 is a coefficient.
3. A *factor* is a number, variable, or sum/differences within a set of parentheses that is being multiplied to something. A *term* is a number, variable, or combination of numerical and variable factors connected with multiplication. $4xy$ is a term with three factors: 4, x, and y.
4. The constant in $4x + 5y(6z - 7) - 2$ is 2. The 7 is part of the factor $6z - 7$, so it does not qualify as a constant.
5. $12e + 8fg + 9h(i + 2j)$ has three terms. $12e$ is the first term with two factors: 12 and e. $8fg$ is the second term with three factors: 8, f, and g. And $9h(i + 2j)$ is a single third term with three factors: 9, h, and $(i + 2j)$.
6. The coefficient of $5x$ is 5 and the coefficient of y is 1, so the sum of the coefficients is 6.

6

Evaluating Algebraic Expressions

STANDARD PREVIEW

In this lesson we will cover Standard 6.EE.A.2.C. You will learn how to evaluate single-variable and multivariable algebraic expressions for the values of their variables. You will see how this skill can be applied to algebraic formulas modeling real-world situations.

Replacing Variables with Number Values

In the last lesson you learned that an algebraic expression is one or more terms, separated by addition or subtraction, where at least one of the terms contains a variable. **Evaluating algebraic expressions** means replacing the variables with real numbers and performing the arithmetic operations in the expression. Although this sounds easy enough, it is critical that you remember to follow the correct order of operations according to PEMDAS. This is an important skill because many real-life situations are modeled using algebraic expressions. Sometimes these expressions are only useful when you are able to plug in real-world values for the variables and then correctly evaluate the expression. This skill will help you get useful information, such as the volume of your pool that needs water, the surface area of your living room that needs to be painted, or the perimeter of a piece of property that needs a fence.

Evaluating Single Variable Expressions

The most basic algebraic expressions contain only one variable and are called *single-variable expressions.* You probably already have some experience working with single-variable expressions, since one example is the expression that represents the perimeter of a square. The *perimeter* of a polygon is the distance around it and the formula for the perimeter of a square is *perimeter* = $4s$, where s represents the length of one of its four congruent sides. If we want to find the perimeter of a square piece of property that has a side length of 62 feet, we replace the s in the expression $4s$ with the value of 62:

Perimeter = $4s$, evaluate at $s = 62$ feet
Perimeter = $4(62) = 248$ feet

Often, you will learn to evaluate expressions through practice problems that don't have any real-world context. Here's a common type of problem for evaluating expressions:

Example: *What is the value of the expression* $3(4x + 50)$ *when* $x = -10$*?*
Solution: Replace the x with -10 and follow the appropriate order of operations mapped out in PEMDAS:

- $3(4(-10) + 50)$; do the multiplication inside the parentheses first
- $3(-40 + 50)$; next do the addition inside the parentheses
- $3(10)$; multiply
- 30 is your final answer

Sometimes there will be a single variable, but it will appear more than once. In this case you will need to replace the variable with its numerical equivalent every time the variable appears before using PEMDAS:

Example: *Evaluate the expression* $-3x^2 + 10x$ *for* $x = -5$.
Solution: Replace both x variables with –5 and follow the appropriate order of operations as laid out in PEMDAS:

$-3(-5)^2 + 10(-5)$; do the exponent first: $-5 \times -5 = 25$
$-3(25) + 10(-5)$; next do the multiplication
$-75 + -50$; add
-125 is your final answer

Subbing Negative Values in for Variables

Did you notice above that when you substituted the –5 in for x^2, the negative sign canceled out when $(-5)^2$ was performed? It's critical to understand that when substituting a negative value in for a variable, use parentheses and include the negative sign in the operation instructed by the exponent. Therefore, *even exponents* will always cancel out the negative sign of a negative base since every product of two negative factors is positive. Conversely, all odd exponents will preserve the negative sign of a negative base since after all the multiplication, there will be one negative factor left over.

Even exponents will cancel out the sign of a negative base: $(-3)^2 = (-3)(-3) = 9$

Odd exponents will preserve the sign of a negative base: $(-2)^3 = (-2)(-2)(-2) = -8$

ERROR ALERT! A common mistake students make is using a negative coefficient to cancel out a negative base *before* acting upon the exponent. This is an especially easy mistake to make when the coefficient is –1! For example, when evaluating $-x^2$ for $x = -4$, it is required that you do *negative four squared* first, before multiplying it by the –1 coefficient:

Example: *Evaluate* $-x^2$ *for* $x = -4$
Solution: The coefficient here is –1 and that will get multiplied by x^2 *after* the exponent is done: $-1(-4)^2 = -1(16) = -16$. Notice that even though the exponent is even, the answer is negative since the coefficient is negative. Watch out for this common mistake!

Practice 1

1. What is the value of $7t - 8$ when $t = 3$?
2. What is the value of $40 \div p \times 2$ when $p = 10$?
3. Evaluate $\frac{-10 + 3c}{4c}$ for $c = -2$
4. What is the value of the expression $5h^2 - 10h^4$ when $h = 2$?
5. What is the value of the expression $5h^2 - 10h^4$ when $h = -2$? How does this answer compare with your answer to question 4?
6. Let w represent any real number other than zero. Will the value of the expression $-w^2$ sometimes, always, or never be negative? Explain your reasoning.
7. Let v represent any real number other than zero. Will the value of the expression $-v^3$ sometimes, always, or never be negative? Explain your reasoning.
8. Marco and Polo are making props for the school play. The prop is a cube that will be used as a pulpit. Marco wants to make a cube that has a side length of two feet and Polo thinks it would be better to make a cube with a side length of three feet. How much bigger is the surface area of Polo's cube than Marco's cube? (Use the formula for the surface area of a cube, *surface area* = $6s^2$, where s is the side length of the cube.)

Evaluating Multivariable Expressions

Although the formulas for area, surface area, and volume of cubes have just one variable for side length, many algebraic expressions contain multiple variables. These types of expressions are called multivariable expressions. If you've ever calculated the *perimeter* of a rectangle using the formula *perimeter* = $2l + 2w$, then you already have some experience with multivariable expressions. To evaluate a multivariable expression, replace each variable in the expression with the given value for that variable. As you did with single variable expressions, remember to be careful as you work through the order of operations.

Example: *Evaluate the expression* $-3m - 10n$ *for* $m = -5$ *and* $n = -2$
Solution: Replace m with -5 and n with -2 and let PEMDAS be your guide as you work through your operations:

- $-3(-5) - 10(-2)$; do the multiplication first
- $15 - (-20)$; subtracting a negative is the same as adding a positive
- $15 + 20$; add to get 35 as your final answer

Now lets work through an example of a real-world multivariable expression that you can evaluate to get useful information:

Example: *Rikki is making a chicken coop along the edge of her property. She constructed a wooden frame that is 18 feet long by five feet wide. She knows that the formula for perimeter is* $2l + 2w$, *where* l *is the length in feet and* w *is the width in feet, and can be used to find the distance around her coop. Evaluate the perimeter formula for* $l = 18$ *and* $w = 5$ *in order to find out how many linear feet of chicken wire Rikki must buy in order to wrap around the frame.*

Solution: Using *perimeter* $= 2l + 2w$, replace l with 18 and w with 5 and work through your operations following PEMDAS.

- $2(18) + 2(5)$; do the multiplication first
- $36 + 10$; add to get 46 as your final answer
- Rikki needs to buy 46 feet of chicken wire to enclose her coop

Practice 2

1. Write an expression to model *15 more than twice x*. Evaluate it for $x = 3.5$
2. Write an expression to model *w is tripled and reduced by 10*. Evaluate it for $w = 9$
3. Evaluate the expression $\frac{1}{2}bh$ for $b = 14$ and $h = 10$. If you recognize this formula, explain what it is used for and what the variables represent.
4. What is the value of the expression $3.14r^2h$ at $r = 10$ and $h = 2$? If you recognize this formula, explain what it is used for and what the variables represent.
5. Evaluate the expression $2(lw + wh + hl)$ for $l = 6$, $w = 4$, $h = 2$. If you recognize this formula, explain what it is used for and what the variables represent.
6. Interest is the money an investment pays you. One formula used to calculate the new amount of your money including interest is $C(1 + r)^t$, where C stands for your beginning investment, r is the interest rate as a decimal, and t is the time in years. What is the value of the expression $C(1 + r)^t$ for $C = 10{,}000$, $r = 5\%$, and $t = 1$? (*Hint:* Don't forget to change your interest rate of 5% into a decimal.) What will the value of your investment be after four years?

Answers

Practice 1

1. $7(3) - 8 = 21 - 8 = 13$
2. $40 \div 10 \times 2 = 4 \times 2 = 8$
3. $\frac{-10 + 3(-2)}{4(-2)} = \frac{-10 + -6}{-8} = 2$
4. $5(2)^2 - 10(2)^4 = 20 - 160 = -140$
5. $5(-2)^2 - 10(-2)^4 = -140$; this was the same answer as question 4 because the negative signs were canceled out by the even exponents.
6. The expression $-w^2$ will always be negative for non–zero values of w. The even exponent will guarantee that w^2 will be positive for all non–zero values of w, but the coefficient of –1 will always make $-w^2$ negative.
7. Sometimes. The expression $-v^3$ will be negative when v is positive. The expression $-v^3$ will be positive when v is negative.
8. Marco's cube: $6s^2$ for $s = 2$: $6(2)^2 = 6(4) = 24$ square feet.
   Polo's cube: $6s^2$ for $s = 3$: $6(3)^2 = 6(9) = 54$ square feet. Polo's cube would have a surface area 30 square feet bigger than Marco's cube. (It would be more than double.)

Practice 2

1. $15 + 2x$ at $x = 3.5$ will equal 22
2. $3w - 10$ at $w = 9$ will equal 17
3. $\frac{1}{2}bh$ at $b = 14$ and $h = 10$ will equal 70. This formula is for the area of a triangle where b = base and h = height.
4. $3.14r^2h$ at $r = 10$ and $h = 2$ will be $3.14(10)^2(2) = 3.14(100)(2) = 628$. This formula is for the volume of a cylinder where r = radius and h = height.
5. $2[6(4) + 4(2) + 2(6)] = 88$ This formula is for the surface area of a rectangular prism where l = length, w = width, and h = height.
6. $10{,}000(1 + 0.05)^1 = 10{,}500$
   $10{,}000(1 + 0.05)^4 = 12{,}155.06$

7

Writing Equivalent Expressions

STANDARD PREVIEW

In this lesson we will cover **Standards 6.EE.A.3** and **6.EE.A.4**. You will first learn how to create equivalent expressions by applying the properties of operations to algebraic expressions. You will then learn how to simplify expressions by combining like terms.

What Are Equivalent Expressions?

Two numerical expressions are **equivalent** if they always have the same value. For example, the numerical expressions $3 + 2$ and $4 + 1$ are equivalent because both sums are equal to 5. When two expressions are equivalent, we can make an equation like $3 + 2 = 4 + 1$.

When variables are involved, we are no longer working with numerical expressions, but instead are dealing with algebraic expressions. Two algebraic expressions are only equivalent if they are always equal, regardless of the value that is chosen for the variable. *If the same value is plugged into both expressions and each expression yields the same result, then two expressions are said to be equivalent.*

Examples of Equivalent Expressions

1. x is equivalent to $x + 0$, because no matter what value x has, adding 0 to it does not change its value. So $x = x + 0$.
2. a is equivalent to $1a$. Multiplying any number by 1 does not change the value of the number. So $a = 1a$.
3. $5 + y$ is equivalent to $2 + y + 3$. No matter what variable is plugged in for y it will always be true that $5 + y = 2 + y + 3$.

The Properties of Operations

You are already familiar with the properties of operations that define equivalent expressions with numbers. For example, you know that $5 + 10$ is the same thing as $10 + 5$ and that $2 \times 3 = 3 \times 2$. These are both examples of the Commutative Property, which states that the order of the numbers doesn't matter with multiplication and addition. You probably remember that the Commutative Property does not work for subtraction and division, since $8 - 2 \neq 2 - 8$ and $12 \div 4 \neq 4 \div 12$. All the properties that you have applied to numbers in the past *also* hold true and apply to algebraic expressions. Let's review these properties so that you can see how we will use each property to help us simplify algebraic expressions.

The Commutative Property

The **Commutative Property** states that in addition and multiplication the order of the numbers does not matter: $4 + 7 = 7 + 4$ and $3 \times 2 = 2 \times 3$. To generalize, we can say that for any numbers a and b:

$$a + b = b + a$$
$$ab = ba$$

Example: *Use the Commutative Property to find an expression equivalent to* $2 + x + 3$.

Solution: The Commutative Property allows $2 + x + 3$ to be rewritten as $2 + 3 + x$. Then add the constants to get $5 + x$. The expression $5 + x$ is equivalent to $2 + x + 3$ for all values of x.

Example: *Use the Commutative Property to find an expression equivalent to* $2v \times 4w$.

Solution: Since a coefficient next to a variable, like $2v$, is the same multiplication, $2 \times v$, $2v \times 4w$ can be written as $2 \times v \times 4 \times w$. The Commutative Property allows the order of the multiplication to be changed: $2 \times 4 \times v \times w$. After multiplying the constants 2 and 4 we get $8vy$. The expression $2v \times 4w$ is equivalent to $8vw$ for all values of v and w.

The Commutative Property with Subtraction

Although the Commutative Property does not work on subtraction, remember that *keep–switch–switch* can always be used to turn subtraction into addition; *then* the Commutative Property can be applied. It is irrelevant if the sign of the terms themselves are negative, as long as the operation connecting the terms is addition.

Example: *Use the Commutative Property to write an equivalent expression to* $6 + x - 10 + y - 5$.

Solution: Use *keep–switch–switch* to change subtraction into addition and then rearrange the terms so that they can be combined:

$6 + x - 10 + y - 5 =$

$6 + x + (-10) + y + (-5) =$

$6 + (-10) + (-5) + x + y =$

$-9 + x + y$

Therefore, $-9 + x + y$ is equivalent to $6 + x - 10 + y - 5$ for all values of x and y.

The Associative Property

The **Associative Property** states that when adding or multiplying three numbers, it does not matter which two you operate on first. For example, when adding 5 + 3 + 13, it makes no difference whether you add 5 + 3 first, and add that sum to 13, or whether you add 3 + 13 first and add the sum to 5.

5 + 3 + 13	5 + 3 + 13
8 + 13	5 + 16
21	21

The same is true of multiplication:

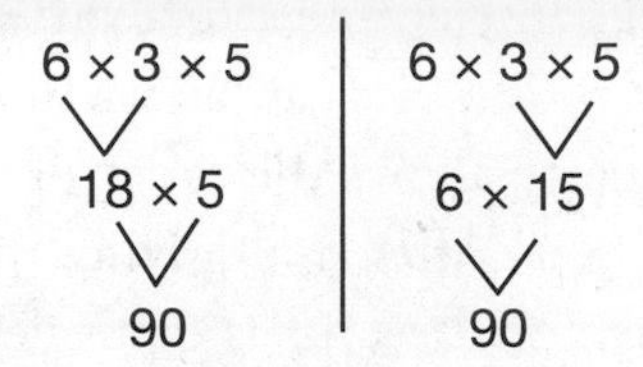

The Associative Property can be generalized for any numbers a, b, and c:

$(a + b) + c = a + (b + c)$
$(ab)c = a(bc)$

In other words, any time we need to add or multiply a bunch of numbers or variables together, we don't need parentheses to group the numbers, since order does not change the sum or product. Like the Commutative Property, the Associative Property does not work on subtraction or division.

Example: *Use the Associative Property to find an expression equivalent to the product* $8(9v)$.
Solution: Here there are three terms being multiplied together: 8, 9, and v. The expression is presented so that the 9 and v are grouped together, but the Associate Property allows for this to be rewritten: $8(9v) = (8 \times 9)v = 72v$.

Example: *Use the Associative and Commutative Properties to find an expression equivalent to* $4(x + 5)$.
Solution: First, remember that multiplication is the same as repeated addition: 4×3 means four groups of three: $3 + 3 + 3 + 3 = 12$. Therefore, $4(x + 5)$ translates to adding $(x + 5)$ to itself 4 times:

$(x + 5) + (x + 5) + (x + 5) + (x + 5)$

The Associative Property states that grouping is insignificant in addition, so we'll remove the parentheses:

$x + 5 + x + 5 + x + 5 + x + 5$

The Commutative Property allows the order of terms being added to be switched:

$x + x + x + x + 5 + 5 + 5 + 5$

Since we have a grouping of four xs we can write that as $4x$, and we can add the four fives to get a final answer of $4x + 20$. The expression $4(x + 5)$ is equivalent to $4x + 20$ for all values of x.

The Distributive Property

The last example we just did leads us nicely into the **Distributive Property**. The Distributive Property distributes a numerical factor to two or more terms that are added or subtracted inside a set of parentheses. For instance, in the expression $5(3 + 6)$, the 5 can be distributed by multiplying it to both the 3 and 6 inside the parentheses and then the two products are added:

$$5(3 + 6) = 5 \times 3 + 5 \times 6 = 15 + 30 = 45$$

The Distributive Property also holds true for algebraic expressions. In variable form, for any numbers a, b, and c: $a(b + c) = ab + ac$.

Example: *Use the Distributive Property to find an expression equivalent to* $4(p + 5)$.
Solution: The Distributive Property calls for the factor outside the parentheses, 4, to be multiplied to both terms inside the parentheses:

$$4(p + 5) = 4 \times p + 4 \times 5 = 4p + 20$$

The expression $4(p + 5)$ is equivalent to $4p + 20$ for all values of x. (Notice that this is the same conclusion that we arrived at using the Associative and Commutative Properties.) We will spend more time with this property in Lesson 14.

Example: *Use the Distributive Property to create an expression equivalent to* $2m + 5m$.
Solution: First notice that $2m + 5m$ is the same thing as $2 \times m + 5 \times m$ and the greatest common factor of $2 \times m$ and $5 \times m$ is m. The Distributive Property can be reversed by pulling the m common factor to the *outside* of a set of parentheses: $2m + 5m = m(2 + 5)$. Add the terms inside the parentheses to get $m(7)$, which can be rewritten as $7m$, according to the Commutative Property. (It is a standard convention in algebra to write the coefficient *before* the variable in a term.)

Practice 1

1. Is the expression $4 + (-x)$ sometimes, always, or never equivalent to the expression $(-x) + 4$? Justify your answer by using a mathematical property.

2. Is the expression $12 \div r$ sometimes, always, or never equivalent to the expression $r \div 12$? Justify your answer by using a mathematical property.

3. Use only the Associative and Commutative Properties to find an expression equivalent to $3(x + 2) + 4(x - 6)$.

4. Write an expression equivalent to $(3e)(4f)(-5g)$.

5. Write three different expressions that are equivalent to $2(4 + m)$ and justify your answers.

6. Use the Distributive Property to show that $2w + 3w + 4w = 9w$.

7. Explain why the expressions $b + 5$ and $c + 5$ *are* or *are not* equivalent.

8. Which of the following expressions are equivalent to $24v - 10$?
   $6(4v - 10)$
   $20 + 24v - 10$
   $4(6v - 10) + 30$
   $4(2v + 2v + 2v) - 10$
   $(12v - 5)2$
   $10 - 24v$

9. Use the values of $q = 1$, $q = -1$, and $q = 2$ to determine if $\frac{8q}{q}$ and $8q^2$ are equivalent expressions. Discuss what you notice.

Combining Like Terms

One way to generate two equivalent expressions is by combining like terms. Recall that **terms** are parts of an expression that are being added to or subtracted from each other. An expression may contain many different kinds of terms, and it is important to have a solid understanding of what like terms are so that you can recognize them. **Like terms** are terms that have all the same exact variable or variables.

Examples

1. x and $3x$ are like terms because they have the same variable.
2. $4k$ and $4j$ are not like terms because they have different variables.
3. The expression $2k + 5p - 3k + 6$ has two like terms: $2k$ and $-3k$. Since $5p$ has a different variable and 6 has no variable, the other two terms are unlike terms.
4. All of the terms in the expression $28 + 13.25 + (-4) + \frac{2}{5}$ are like terms, because they are all constants and do not have a variable.
5. $5xy$ and $5x$ and $5y$ are all unlike terms because terms must have *all* the same variables in order to be like. Therefore, $5xy$ and $6xy$ would be like terms.

Combining like terms means adding or subtracting the like terms in an expression. Some expressions are long and messy, and combining the like terms is therefore also referred to as **simplifying an expression**. Think of like terms as being similar types of items that are strewn around your room when you are asked to clean it. All the dirty laundry gets put together and goes in the hamper. All the pens and papers get put together and go in your backpack. All the wrappers from granola bars, gum, and candy go into the trash bin. Combining like terms is like tidying up an algebraic expression! Since you wouldn't put dirty socks on your desk with your pens, or trash in your laundry bin, make sure that when combing variable expressions, keep terms that are not like completely separate.

Let's start with one of the previous examples to see if we can uncover a general rule to use for combining like terms. We stated earlier that x and $3x$ are like terms. Remembering that x is equivalent to $1x$, you could use the Distributive Property to rewrite $1x + 3x$ as $x(1 + 3)$, which then becomes $4x$. After applying the Commutative Property your final answer can be written as $1x + 3x = 4x$. This is a lot of work to go through each time, so we think you'll instead want to use the shortcut of adding and subtracting the coefficients of like terms and keeping the variables the same.

Like terms are terms that have the exact same variable or variables. In order to add or subtract like terms, add or subtract the coefficients and keep the variables the same. Ex: $10x + 8x = 18x$. Since $8y$ and $8x$ are *not* like terms, the expression $8y + 8x$ cannot be combined to form an equivalent expression $16xy$. $8y + 8x \neq 16yx$.

Watch how like terms are combined in the following examples:

Examples

$7p + 5p$ is equivalent to $12p$. $7p + 5p = 12p$

$14b - 9a + 10b$ is equivalent to $24b - 9a$. Since a and b are different variables, their terms cannot be combined. $14b - 9a + 10b = 24b - 9a$.

ERROR ALERT! Although most students tend to not have a problem understanding that $5x - 2x = 3x$, many get confused when faced with just the term "x." The coefficient of a lone variable is always "1" and it helps to write in the 1 before adding or subtracting like terms so that you're not tempted to cancel out the variables: $5x - x = 5x - 1x = 4x$.

Practice 2

1. Which of the following expressions have like terms that can be combined?

   $4(x + y)$
   $4x + y$
   $5x + 5y$
   $4xy$
   $4x + 5y + x$
   $4x + 5$

2. If the Commutative Property does not hold true for subtraction, then why can $7x - 5y - 20x - 10y$ be simplified as $-13x + (-15y)$? Show the steps of how this is done and explain your reasoning.

3. Simplify the expression $L + W + L + W$.

4. While doing his homework, Hudson wrote that $2x + 4y = 6xy$. Ryder noticed Hudson's answer and told him that $2x$ and $4y$ cannot be combined to make $6xy$. Hudson refuses to believe Ryder. Help Ryder argue his case! Choose two different values for x and y and write them here: $x =$ ___ and $y =$ ___. Now use those two values to evaluate $2x + 4y =$ ________________ and $6xy =$ ______. What do you notice?

5. Given the expression $5(2 + x) + 3(4 + x)$, Noah thinks that the principle of multiplication must be used to expand $5(2 + x)$ and $3(4 + x)$ before combining the like terms. Tucker thinks he's wrong and it's possible to add the 2 and 4, and the x and x together before simplifying. Explain how this problem must be completed and justify your reasoning using PEMDAS.

Answers

Practice 1

1. The expression $4 + (-x)$ is always equivalent to the expression $(-x) + 4$ since the Commutative Property states that order of terms is not significant in addition.
2. The expression $12 \div r$ is never equivalent to $r \div 12$ since the Commutative Property does not apply to division or subtraction.
3. $3(x + 2)$ is three groupings of $(x + 2)$: $(x + 2) + (x + 2) + (x + 2)$. $4(x - 6)$ is four groupings of $(x - 6)$: $(x - 6) + (x - 6) + (x - 6) + (x - 6)$. After the Associative Property removes all the parentheses and the Commutative Property rearranges the terms, the sum of $3(x + 2) + 4(x - 6)$ can be written as $7x + -12$.
4. Use the Associative and Commutative Properties to rewrite $(3e)(4f)(-5g)$ as $3 \times 4 \times -5 \times e \times f \times g = -60 \times efg = -60efg$.
5. $2(4 + m)$ is equivalent to $4 + m + 4 + m$. It is also equivalent to $8 + 2m$. Any other expression where the constants add up to 8 and the variables add up to $2m$ would be an equivalent expression (like $7 + 1 + m + m$).
6. The Distributive Property can be reversed to pull the factor of w out of $2w + 3w + 4w$: $w(2 + 3 + 4)$. Then the constants inside the parentheses can be combined to get $w(9)$ or $9w$.
7. $b + 5$ and $c + 5$ are not equivalent because b and c could be different values and do not have to be the same value.
8. The following expressions are equivalent to $24v - 10$:
   $4(6v - 10) + 30$
   $4(2v + 2v + 2v) - 10$
   $(12v - 5)^2$
9. $\frac{8q}{q}$ and $8q^2$ equal the same values when $q = 1$ and $q = -1$, but when $q = 2$ the two expressions do not return the same answer, so they cannot be equivalent expressions.

Practice 2

1. Only $4x + 5y + x$ has like terms that can be combined: $4x + 5y + x = 5x + 5y$
2. The Commutative Property does not hold true for subtraction, but after *keep–switch–switch* has been used to rewrite the problem as addition, the terms can be rearranged and combined as follows:
   $7x - 5y - 20x - 10y =$
   $7x + -5y + -20x + -10y =$
   $7x + -20x + -5y + -10y =$
   $-13x + -15y$

3. $L + W + L + W = 2L + 2W$
4. We will choose $x = 0$ and $y = 5$ (but any two different numbers could have been chosen)

   $2x + 4y = 2(0) + 4(5) = 20$

   $6xy = 6(0)(5) = 0$

   Ryder is correct that $2x$ and $4y$ cannot be combined to make $6xy$
5. PEMDAS states that multiplication must come before addition, so the 5 and 3 must be distributed to both terms inside their adjoining parentheses before any like terms can be combined. Noah was correct and Tucker was mistaken.

8

Testing and Writing Equations and Inequalities

STANDARD PREVIEW

In this lesson we will cover Standards 6.EE.B.5 and 6.EE.B.6. You will learn how to use substitution to determine if a number is a solution to an equation or inequality and how to use variables to represent real-world situations. You will also learn to carefully consider what kind of numbers make sensible input values in real-world algebraic expressions.

When Are Equations "True"?

We know at some point you may have lied about your age or height. You may have wanted to get into a PG–13 movie when you were only 12 years old, or maybe you said you were two inches taller than you really were so you could get on the best roller coaster at your favorite theme park. We forgive you for these lies, but don't make it a habit! And don't be surprised when we tell you that equations lie too sometimes. In fact, it's pretty easy to see that the numerical equation $5 = 3 + 4$ is lying right to our faces! The sooner you learn this the better: just because an equation has an equals sign, don't trust that it's telling the truth. Since this is a math book and not a courtroom, we don't say that $5 = 3 + 4$ is a liar; but we do say that $5 = 3 + 4$ is *not a valid equation*, or that it is *false*. Conversely, we say that an equation like $5 = 3 + 2$ is *valid*, or *true*.

Examples

1. The equation $10 = 4 + 6$ is true.
2. The equation $52 = 10$ is false.
3. The equation $5{,}999{,}999 \times 0 = 0 \times (-485)$ is valid.
4. The equation $-(4)^2 = 16$ is invalid.

Testing Equations

An **algebraic equation** is an equation that has one or more variables. (Algebraic *expressions* don't have an equals sign, algebraic *equations* do.) Algebraic equations can be pretty sly in their deception and it takes a little more work to see if they are *valid* or *invalid*. We'll start with a simple equation, $x + 3 = 4$. Whether this is a true or false equation depends on the value of x. The equation is *true* when $x = 1$, but it is *false* when x equals all other values.

In order to test if an equation is true for a given value, plug that value in for the variable, follow the correct order of operations to evaluate the numerical expressions on each side of the equals sign, and see if the resulting numerical statement is true or false. Let's try some:

Example: *Is* $2x = 16$ *true for* $x = 8$?
Solution: $2x = 16$ is true for $x = 8$ since $2(8) = 16$.

Example: *Is the equation* $20 = 50 - 3x$ *true for* $x = 1$?
Solution: Plug 1 in for x: $20 \neq 50 - 3(1)$, so this equation is false for $x = 1$.

Example: *Is* $20 = 50 - 3x$ *true for* $x = 10$?
Solution: Plug 10 in for x: $20 = 50 - 3(10) = 20$, so this equation is true for $x = 10$.

When Are Inequalities "True"?

What's the difference between your friend saying, "Meet me at my house at 10:00 A.M.," versus "Come by my house sometime after 10 A.M.?" Preciseness is the difference! The first statement gives an exact time, while the second statement has multiple solutions. Given the first statement, your friend might get angry if you show up at 10:30. However, if she had said *come by my house sometime after 10 A.M.*, she probably wouldn't mind if you show up at 11:00 A.M. This varying degree of preciseness is found between equations and inequalities as well. The equation $t = 10$ represents *meet me at my house at 10:00 A.M.*, while the inequality $t > 10$ represents *come by my house sometime after 10 A.M.* Notice that the equation $t = 10$ has only one value for t that makes it true: 10. Meanwhile, the inequality, $t > 10$, has lots of potential solutions that would make it true: $t = 10.5$, $t = 10.75$, and $t = 12$.

ERROR ALERT! Students forget that inequalities shouldn't always be read left to write. Reading $-5 > x$ as *negative 5 is greater than* x sounds so confusing! Instead, begin with the variable and say "is less than" if the symbol is pointing *toward* the variable and say "is greater than" if the symbol is pointing *away* from the variable. If the symbol is underlined, think of that as half of an equals sign and include "or equal to" in your translation of it:

- $7 < m$: *m is greater than seven*
- $v \geq 12$: *v is greater than or equal to 12*
- $k < 42$: *k is less than 42*
- $10 \geq a$: *a is less than or equal to 10*

In order to test if an inequality is true or false, follow the same steps used in equations:

Example: *Is* $2x \geq 16$ *true for* $x = 8$*?*
Solution: $2x \geq 16$ is true for $x = 8$ since $2(8) = 16$ and $16 \geq 16$.

Example: *Is* $3y > 24$ *true for* $x = 8$*?*
Solution: $3y > 24$ is false for $x = 8$ since $3(8)$ is equal to 24 and not greater than 24

Example: *Is the inequality* $20 < 50 - 3x$ *true for* $x = 1$?
Solution: Plug 1 in for x: $20 < 50 - 3(1)$, so it is true for $x = 1$.

Example: *Is the inequality* $20 \geq 50 - 3x$, *true for* $x = 10$?
Solution: Plug 10 in for x: since $50 - 3(10) = 20$ and $20 \geq 20$, that means the inequality is true for $x = 10$.

Practice 1

1. Is the equation $9 - 3^2 = 0$ true?

2. Is the equation $p^3 - 1 = 7$ true for $p = 2$?

3. Is the equation $c^2 + 4 = 12$ true for $c = 4$?

4. Is the inequality $40 - 8b > 0$ true for $b = 6$?

5. Is the inequality $5r \leq 30 - 10r$ true for $r = 2$?

6. Will the equation $x + 1 = x + 500$ ever be true for any value of x? Explain your reasoning.

7. Is the equation $6g^2 = 54$ true for $g = 3$?

8. Can you think of another value for g that would make the equation in question 7 true? Explain your reasoning.

Variables in the Real World

In Lesson 5 you practiced translating words into mathematical expressions. Now that you are getting familiar with variables, we are going to use variables to represent real-world mathematical expressions and equations. Furthermore, we're going to consider the context of the situation to determine the *types* of numbers that make sensible inputs for the variables in our expressions.

Example: Molly had \$38 with her when she went to the movies. She bought a movie ticket for \$12 and she bought m boxes of Milk Duds that cost \$3 each. Represent Molly's money using an algebraic expression, but do not combine any like terms. What kind of numbers would work in the expression for m?

Solution: Since Molly started with \$38 and bought a \$12 ticket, begin with \$38 – \$12. Since Molly bought m boxes of Milk Duds at \$3 each, her Milk Duds cost will be $\$3 \times m$ or $3m$. (Remember that the word "each" is a hint that multiplication is being used.) Subtract $\$3m$ from the first part of the expression to get $\$38 - \$12 - \$3m$. Inputs for m must be whole numbers since Molly can't buy part of a box of Milk Duds. m must also be small enough so that when the expression is evaluated it is still a positive number. This is because once the value of the expression is less than zero, that means that Molly has spent all her money and maybe had to borrow money from a stranger in line at the concession stand.

Sometimes you will be given enough information to write an equation instead of just words. Look for numerical or variable information to be associated with verbs like "is," "earns," and "makes" since verbs indicate that it's time to write an equals sign!

Example: Daniel brings metal cans and glass bottles that he collects on the street to a recycling center in order to earn money for food. The recycling center pays him five cents per can and four cents per pound of glass. If Daniel collects c cans and p pounds of glass on a certain day and earns \$32.60. Write an equation representing his can and bottle recycling on that particular day. What types of numbers would make sense for c and p?

Solution: Since the word "per" indicates multiplication, *five cents per c cans* will be written as $\$0.05c$. *Four cents for p pounds* should be written as $\$0.04p$. Add these two terms together to get $\$0.05c + \$0.04p$ and set this equal to \$32.60 since that is the total amount of money Daniel earned: $\$0.05c + \$0.04p = \$32.60$. Since Daniel cannot get money for partial cans, c must be a whole number greater than or equal to zero. However, since it is possible for Daniel to recycle partial pounds of glass, p can be any fractional or decimal number greater than or equal to zero. (When fractional and decimal answers are acceptable, it is commonly said that *p can be any real number greater than zero.*)

Finding Appropriate Values for Variables

Before we let you practice this on your own, let's further discuss what types of numbers the variables in expressions or equations could be. In general, you first want to consider if it makes sense to plug in only **whole numbers** (as in *cans* being recycled) or if it is reasonable to plug in **fractional numbers** (as in *pounds* of glass being recycled). That will depend on the context of the question: the same item might only be represented with whole numbers in one context, but fractional numbers might work for that same item in a different context. (For example, James can only *buy* whole slices of pizza, but James can *eat* fractional slices of pizza.)

Next you want to consider if only **positive numbers** can be used or if it's okay to include **negative numbers**. A shop cannot have a negative number of customer purchases for the month, but they *can* make a negative amount of money if their expenses are greater than their income. Although temperature can be negative, the number of inches of rainfall per year cannot be negative.

Lastly, you want to look for any restrictions required by **special circumstances**. For example, if a variable is standing for the number of the months, it will have to be between 1 and 12, but if the variable represents the day of the month, it would be appropriate to consider numbers between 1 and 31. For the most part, this skill takes a little curiosity combined with some common sense. Ask yourself, "What am I really solving for here? What kind of numbers make sense in the real-world context of this problem?"

Practice 2

1. Cleo has a sales booth in a mall where she sells holiday candles that cost $12. Her profit for December when she sells c candles is modeled by the expression $\$9c - \$1{,}500$. What kind of numbers could c be? Do you have any ideas what the $1,500 might represent?

2. On a Spanish test, the spelling questions are worth 5 points, the vocabulary questions are worth 8 points, and the verb conjugation questions are worth 10 points. If Miss Madeleine does not give any partial credit, what expression will she use to calculate the total point score for her students who get s spelling questions, v vocabulary questions, and c verb conjugations correct? What kinds of numbers make sense for each of these variables?

3. For every Sonos speaker Free My Music sells, Matt earns $\frac{1}{4}$ of the price of the speaker. When he goes to professionally install speakers in a client's home he earns $100 per installation. Write an expression that models Matt's January income if he sold s dollars in speakers and did h home installations. What types of numbers would make sense for s and h?

4. Monica averages 62 miles per hour for t hours. In total she drove 289 miles. Write an equation that represents this situation and discuss what kinds of numbers t could be.

5. Cole has a strange but consistent technique for choosing her "lucky" four numbers when she plays the lottery each month. She determines the first and fourth number by subtracting 2 from the number of letters in the name of that month. She always uses 13 for her lucky middle two numbers. Write an expression modeling how Cole determines her first and last numbers, based on the number of letters in the month, m. What types of numbers will m be?

Answers

Practice 1

1. Yes, $9 - 3^2 = 9 - 9 = 0$
2. Yes, when $p = 2$, $p^3 - 1 = 8 - 1 = 7$
3. No, when $c = 4$, $c^2 + 4 = 4^2 + 4 \neq 12$.
4. No, when $b = 6$, $40 - 8b$ is less than 0 and not greater than 0.
5. Yes, when $r = 2$, $5r < 30 - 10r$ is equivalent to $10 < 30 - 10(2)$, which is true.
6. $x + 1 = x + 500$ will never be true for any value of x because it is impossible to arrive at the same answer when you add 1 as when you add 500.
7. When $g = 3$, $6g^2 = 54$ is equivalent to $6(3^2) = 54$, which is a true equation.
8. -3 would also make $6g^2 = 54$ a true statement since the negative sign will cancel out when g is squared.

Practice 2

1. Since Cleo cannot sell partial candles or negative candles, c must be a whole number greater than or equal to 0. Since the $1,500 is being subtracted, it represents some combination of expenses, which might include (but are not limited to) rent, inventory, employees, and/or insurance.
2. Total Points = $5s + 8v + 10c$. Since Miss Madeleine doesn't give any partial credit, s, v, and c must be whole numbers greater than or equal to zero.
3. January Income = $\frac{1}{4}s + \$100h$. Since his home installations are billed at a flat rate of $100 each, h must be a whole number greater than or equal to 0. Since s is the dollar amount of speaker sold, s does not need to be a whole number and just needs to be greater than or equal to zero.
4. $62t = 289$ miles. t represents hours, so it can be in fractional or decimal format.
5. $m - 2$ represents what number Cole will use first and last on her lottery ticket. Since May has only three letters and September has nine letters, m will be between 3 and 9.

9

Solving One-Step Equations

STANDARD PREVIEW

In this lesson we will cover Standards 6.EE.B.7 and 6.EE.B.8. You will learn how to solve algebraic equations and how to write and graph inequalities that represent real-world conditions.

Single-Step Algebraic Equations

In an earlier lesson we learned how to determine if an equation was true for a given value. That was a warm-up for solving algebraic equations. The phrase *solve an algebraic equation* means finding the value of the variable that makes the equation true. Many students begin to sweat when they see algebraic equations like $9 + x = 10$ because they think that math all of a sudden becomes impossible, but they have actually been solving equations similar to this for years. Instead of having variables, the equations you have worked with probably had blank lines or boxes where you filled in the missing information:

$9 + \_\_\_\_ = 10$

$9 + \square = 10$

So don't panic! There's no need for you to get nervous since you've been doing algebraic reasoning for years!

Solving with Opposite Operations

What operation do you use to solve the question 23 + ____ = 30? Are you adding 23 to 30 or are you subtracting 23 from 30? If you are subtracting 23 from 30 then you are already on your way to algebraic success! **Opposite operations** are used to solve algebraic equations. Although that might be kind of obvious to you in the equation 23 + ____ = 30, it's helpful to have a firm rule to rely on when working with more challenging questions like $\frac{1}{8} + x = 0.35$, so let's establish some ground rules for solving algebraic equations.

How to Solve a One-Step Equation

1. The goal in solving one-step equations is to get the variable alone on one side of the equals sign and have a numerical value on the other side. (We imagine that x is having a party and all of a sudden gets a terrible headache and wants to be left alone. x has to *uninvite* guests by using opposite operations.)
2. See what is "bothering" x; what is on the same side of the equals sign as x? What operation is it using to interact with x?
3. Use the opposite operation to remove the number that is next to the variable. (Although x had initially *asked* that number to come to her party, now x is doing the opposite by asking that number to leave.)
   - Addition and subtraction are opposite operations
   - Multiplication and division are opposite operations
4. The golden rule when working with algebraic equations is that in order to keep the equation balanced, whatever you do to the equation on one side of the equals sign, you must do to the equation on the *other* side of the equals sign. (Think of the equation as being two separate scales that have equal weights on them—if you take one of the weights off the first scale, in order for the second scale to have equal weight, you need to remove the same amount of weight from the second scale.)

Example: *Find the value of x that makes $\frac{1}{8} + x = 0.35$ true.*
Solution:

$\frac{1}{8} + x = 0.35$	(x is being annoyed by $\frac{1}{8}$, so get rid of $\frac{1}{8}$ with opposite operations.)
$-\frac{1}{8} \quad = -\frac{1}{8}$	(Subtract $\frac{1}{8}$ from both sides of the equation to get x alone.)
$x = 0.35 - 0.125$	(Change $\frac{1}{8}$ into a decimal before performing subtraction.)
$x = 0.225$	(This is your final answer.)

Although it's probably not difficult for you to solve the equation $5x = 30$, you can see from the previous example that things get a little more challenging when fractions come to town! Don't panic though, just follow the steps above and things will work out fine. (Remember that dividing by a fraction is the same thing as multiplying by its reciprocal.) Notice in this next question that sometimes your variable will be on the right side of your equals sign—don't let that throw you for a loop!

Example: What value of x makes this equation true? $\frac{2}{3} = \frac{1}{12}x$
Solution:

$\frac{2}{3} = \frac{1}{12}x$	(x is being annoyed by $\frac{1}{12}$, so get rid of $\frac{1}{12}$ with opposite operations.)
$\div \frac{1}{12} \quad = \div \frac{1}{12}$	(Divide both sides of the equation by $\frac{1}{12}$ to get x alone.)
$\frac{2}{3} \div \frac{1}{12} = \frac{2}{3} \times \frac{12}{1}$	(Divide by $\frac{1}{12}$ by multiplying by its reciprocal.)
$x = 8$	(This is your final answer.)

Setting Up and Solving Real-World Problems

Let's take a look at how we could represent a word problem with an algebraic equation and then solve it to find our answer. After all, this is what algebra is all about—modeling real-world situations with equations so that we can determine solutions!

Example: Mr. B. has 36 pounds of cactus soil and he wants to raise money for the school garden by selling tiny planters of mixed succulents at his school's GardenPalooza event this spring. He wants to make p planters that each have $\frac{2}{3}$ of a pound of cactus soil. Write and solve an equation to model how many planters Mr. B. can make with his 36 pounds of soil.

Solution: As soon as fractions enter the scene many students throw their hands up in the air and exclaim, "I have no idea!" One trick to dealing with fractions in word problems is to replace them with *nice numbers. Nice numbers* are numbers that are easy to work with, like two. What if Mr. B. had 36 pounds of soil and wanted to make planters with *two pounds of soil each*? You'd probably remember that *each* suggests multiplication, and therefore the equation 2 × (# planters) = 36 could represent this situation. Once you've figured out how to set up your problem with *nice numbers*, replace the *nice numbers* with the fractions in the original question. Since Mr. B has 36 pounds of soil and each p planter will get $\frac{2}{3}$ of a pound of cactus soil, model this with the equation $36 = \frac{2}{3}p$. Now solve it using opposite operations:

$36 = \frac{2}{3}p$	(p is being annoyed by $\frac{2}{3}$, so get rid of $\frac{2}{3}$ with opposite operations)
$\div \frac{2}{3} \quad = \div \frac{2}{3}$	(Divide both sides of the equation by $\frac{2}{3}$ to get p alone.)
$36 \div \frac{2}{3} = 36 \times \frac{3}{2} = p$	(Divide by $\frac{2}{3}$ by multiplying by its reciprocal.)
$p = 54$	

Mr. B will be able to make 54 succulent planters to sell this spring. Get out those gardening gloves!

When solving algebraic operations use opposite operations to get the variable alone.

- **Addition and subtraction are opposite operations**
- **Multiplication and division are opposite operations**

Whatever you do to one side of an equation, you must also do to the other side of the equation in order to keep it balanced.

How to Check Your Solutions

In the previous lesson you were asked to determine if an equation was true for a given value for its variable. In this lesson you are finding the value for the variable that makes the equation true. *Solve* $40x = 20$ is the same way of saying *Find the value of x that makes the equation* $40x = 20$ *true*. Therefore, once you solve any equation for the value of its variable, you can plug that value back into the original equation to make sure it produces a true equation. If it does, then your answer is correct. If it doesn't, then you've made a mistake somewhere along the way and you'll need to recheck your work.

Example: *Izzy and Eli are both trying to solve the equation* $60 = \frac{3}{4}x$. *Eli got* $x = 45$ *and Izzy got* $x = 80$. *Determine who has the correct answer.*
Solution: Plug each of their values in for x in the given equation and see which answer makes the equation true:

- Eli ($x = 45$): $60 = \frac{3}{4}(45)$. This answer yields the equation $60 = 33.75$, which is false, so Eli's answer is incorrect.
- Izzy ($x = 80$): $60 = \frac{3}{4}(80)$. This answer yields the equation $60 = 60$, which is true, so Izzy's answer is correct. Solid job, Izzy!

Practice 1

Find the value of the variable that makes each equation true.

1. $m + 1{,}890 = 4{,}367$

2. $64 = 42f$

3. $\$8.23 = n + \$5.70$

4. $v\frac{1}{100} = 5.7$

5. $\frac{3}{8} + x = \frac{5}{4}$

6. $0.05x = 4.065$

7. $\frac{4}{15} = \frac{2}{9}u$

8. Mr. Hansen marked Tanner's answer incorrect, but Tanner claims he got the right answer. If the equation was $\frac{4}{5}h = 84$ and Tanner's answer was $h = 105$, determine if Mr. Hansen or Tanner was correct.

9. Do all algebraic equations have a solution that makes them true? If not, give an example of an equation that doesn't have a solution.

Inequalities

In the previous lesson we discussed that inequalities have multiple solutions. In life there are many situations that have multiple solutions; therefore, inequalities are very useful in modeling real-world situations. For example, if Juan has more than three cats, the inequality $c > 3$ represents that he could have four, five, or even 20 cats. For all we know, $c > 3$ might represent that Juan has a cat retirement farm in Boyleston with 659 cats (which is unlikely, but possible). The point is, there are an infinite number of solutions that will make every inequality a true statement. This collection of possible answers is called *the solution set* and the solution set to an inequality is commonly graphed on a number line. Before we discuss how to graph inequality solutions sets on number lines, review how to read inequalities in the chart below (this was discussed in Lesson 8):

Example	Read As	Solutions Include
$5 > x$	x is less than five	0, −3, 4.9, but NOT 5
$x \geq 5$	x is greater than or equal to five	5, 6, 70, 347
$5 \geq x$	x is less than or equal to five	5, 4.99, 0, −10
$x > 5$	x is greater than five	6, 7, 100, but NOT 5

Graphing Inequalities

Now we're ready to get graphing, so let's start with the inequality modeling Juan's cats, $c > 3$. Here are the steps for making an inequality number line:

1. Make a number line that has your solution in the middle and counts three units in both directions. (Remember, your numbers get bigger as you move to the right and smaller as you move to the left.)

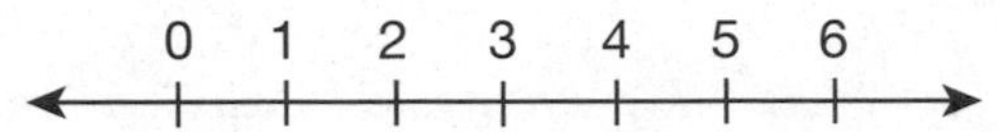

2. This step is important! Circle your numerical solution on the number line. If the symbol in your solution is ≤ or ≥ you need to **shade in your circle** to show that this number is part of the solution set. If the symbol is < or > you need to **keep your circle open** like an "o" to show that this number is *not* part of the solution set:

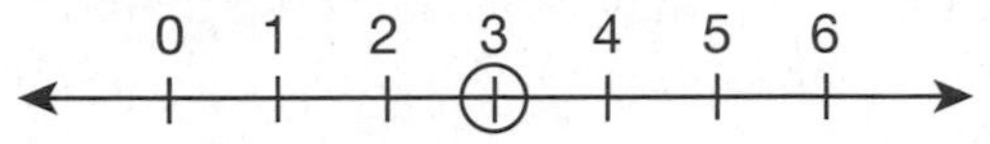

3. The last step is to shade the number line so that it correctly indicates your solution set. If you read your inequality as "greater than" you will shade to the right, and if you read it as "less than" you will shade it to the left. Since Juan has more than 3 cats we shade to the right to demonstrate that:

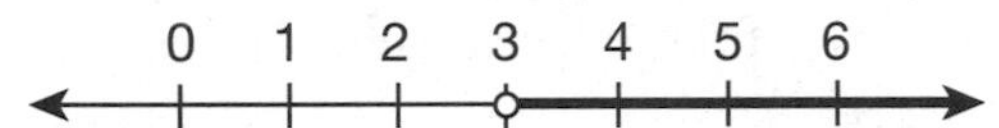

It's as easy as 1-2-3 to graph your solution to an inequality on a number line!

Real-World Inequalities

Since lots of life's problems have multiple solutions, it's important to be able to represent situations using number lines. There are some key words that help us recognize when an inequality is required and which direction the inequality symbol should point! Here's a reference for you to use:

Phrase	Means	Example	Inequality
"at least"	greater than or equal to	Jen has at least three chinchillas	$c \geq 3$
"at most"	less than or equal to	Jen has at most two hedgehogs	$h \leq 2$
". . . than"	will be < or >, not ≤, ≥	It's colder than 10 degrees	$t < 10$
". . . than"	will be < or >, not ≤, ≥	It's hotter than 90 degrees	$t > 90$
"over"	greater than	I spent over 12 hours working	$w > 12$
"under"	less than	The baby weighs under six pounds	$b < 6$

Our chart could go on and on, but hopefully this is enough to give you a sense of how to recognize inequalities and think about their solutions. In general, if you can imagine that there's more than one value that could make sense in a solution, you are dealing with an inequality.

Example: *Summer ran for at least 15 minutes on Monday, 12 minutes on Tuesday, and 18 minutes on Wednesday. Write an inequality to represent Summer's running from Monday to Wednesday and graph this on a number line.*

Solution: Since this problem involves "at least" we know *she ran greater than or equal to* the time given for each day. The total for the 3 days is 45 minutes, but since she may have run more than the stated amount each day, we represent this with the inequality $s \geq 45$. The graph modeling this would look like:

Practice 2

1. How many solutions does each of the equations below have?
   $x = 13$
   $x > 13$
   $x \geq 13$

2. What does it mean for an inequality to have infinite solutions? Why can't we count all the solutions?

3. Since the inequality $x \leq 20$ includes 20, while the inequality $x < 20$ does not, does it make sense to say that $x \leq 20$ has one more solution than $x < 20$?

4. Draw a number line to represent each of the following inequalities:
   a. $m < 99$
   b. $k > 20.5$
   c. $-5\frac{3}{4} \leq v$

5. Write the inequality that models each number line graph:
   a. 12 13 14 15 16 17 18
   b. −6 −5 −4 −3 −2 −1 0

6. Castelle has baked at least 400 cookies this month. Represent this on a number line.

7. Abe is having a birthday party and his mom said he could invite no more than 10 friends. Graph this on a number line.

8. The top of Mount Washington in New Hampshire is the tallest peak in the northeastern United States and records some of the highest wind speeds and lowest temperatures in the world! One day last week the summit had a high temperature of only –12 degrees. Represent this on a number line.

Answers

Practice 1

1. $m = 2{,}477$
2. $f = \frac{64}{42} = \frac{32}{21}$
3. $n = 2.53$
4. $v = 570$
5. $x = \frac{5}{4} - \frac{3}{8} = \frac{7}{8}$
6. $x = 81.3$
7. $u = \frac{4}{15} \div \frac{2}{9} = \frac{4}{15} \times \frac{9}{2} = \frac{6}{5}$
8. $h = 84 \div \frac{4}{5} = \frac{84}{1} \times \frac{5}{4} = 105$, so Tanner's answer was correct.
9. No, not all algebraic equations have a solution that makes them true. The equations $6 + j = 60 + j$ has no solution.

Practice 2

1. $x = 13$ has 1 solution, $x > 13$ has infinite solutions, and $x \geq 13$ has infinite solutions
2. An inequality has infinite solutions in the same way that a ray has a starting point but no end point. Even between the numbers one and two, there are infinite decimals.
3. Although $x \leq 20$ *technically* has one more solution, we still consider $x \leq 20$ and $x < 20$ to have the same number of solutions: infinite. There is no such thing as *infinite plus one*.
4. a.

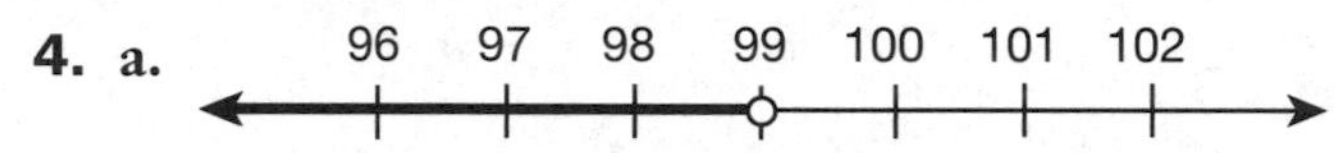

b.

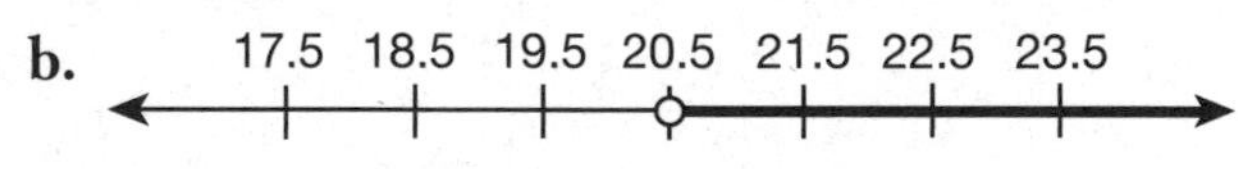

c. −9 −8 −7 −6 −5 −4 −3

5. a. $x \leq 15$

b. $x < -3$

6. 395 396 397 398 399 400 401 402 403 404 405

7. 7 8 9 10 11 12 13

8. −15 −14 −13 −12 −11 −10 −9

10

Modeling with Equations, Tables, and Graphs

STANDARD PREVIEW

In this lesson we will cover Standard 6.EE.C.9. You will learn how to model real-world problems with algebraic equations, tables, and graphs.

Using Two Variables to Model Relationships

Up until now we've spent our time modeling scenarios using just one variable. When an equation has just one variable, like $x + 5 = 8$, there is just one solution that will make the equation true. However, many algebraic models use more than one variable and these variables will change in relation to one another. Sometimes when one variable gets bigger, it causes the second variable to get bigger as well. Other times, as one variable gets larger it causes the second variable to become smaller. Therefore, instead of just one lonely solution, you'll be able to find lots of pairs of solutions that make a two-variable equation true!

Interconnected Variables

In life, few things happen in isolation: whenever something changes, something else usually changes along with it. When your mom drives her car, her fuel gauge will go down. If you babysit more hours, your paycheck will go up. If you buy something at the store, the amount of money in your wallet will go down. If you spend time studying, your grade will hopefully improve. All of these situations involve two variables that influence each other. Each variable represents a real-world factor, and as the value of one variable changes, it influences the other variable to change as well.

> **Example:** *You charge $12 per hour for your babysitting services. Write an equation that represents your pay p in terms of h hours worked.*
> **Solution:** You don't need a fancy algebraic equation to know that as h gets bigger, your pay, p, will also get bigger. Recognizing that *per* indicates multiplication, translate *$12 per hour*, as $\$12 \times h$ and set this equal to your pay: $p = \$12 \times h$.

Looking at this equation you should be able to see that when you work one hour, you will earn $12 and if you work two hours you will earn $24. Therefore, when $h = 1$, $p = 12$ and when $h = 2$, $p = 24$. Basically, the longer you work, the more you will earn. Notice that the number of hours you work *determines* your pay. The variable *h actively changes* and the variable *p reacts* to those changes. Since your pay *depends* on your hours worked, we call it the *dependent variable*. Your hours worked are called the *independent variable* since they are influencing the pay. The relationship between the dependent and independent variable is the foundational relationship that will determine how you set up algebraic equations to model real-world situations.

Dependent and Independent Variables

When two changing quantities (like hours and money) have a specific mathematical relationship, one variable will always be the **independent variable** and the other will be the **dependent variable**. The amount of money you get paid *depends* on the number of hours you work, so *pay* is the dependent variable and *hours* is the independent variable. To determine which variable is independent and which is dependent, just think about which factor is acting and which factor is reacting. Who is bossing who around?

Independent and Dependent Variable Examples

1. The more minutes you spend working out, the more calories you will burn. Since the calories burned *depend* on the minutes, the *minutes* are your independent variable and the *calories burned* are your dependent variable.
2. The more miles you drive your car instead of riding your bike, the more money you will spend on gas. The cost of refueling your car *depends* on how many miles you drove, so *miles* is the independent variable, and cost is the dependent variable.
3. The faster you bike, the less time it will take you to arrive at your destination. The time it takes you to arrive *depends* on how quickly you bike, so your *speed* is the independent variable and *time* is the dependent variable.

Using Context to Determine Dependent and Independent Variables

Notice that in the first two examples, as the independent variable *increased*, the dependent variable also increased. More minutes working out meant more calories burned. More miles driven meant more money spent on fuel. The third scenario had a different kind of relationship—as the independent variable increased, the dependent variable *decreased*; as the speed you biked increased, the time it took to reach your destination *decreased*. It is important to also notice that in this last relationship, *time* was the dependent variable; however, in the example of working out, *time* was the independent variable. The context of the question must always be carefully considered to determine which variable is dependent and which is independent. It is also really important to know that it is most common to focus on how the dependent variable changes as the *independent variable increases*. You will practice this next.

Practice 1

For each question determine which variable is independent and which is dependent. Then determine if the dependent variable increases or decreases as the independent variable increases. Fill in the blanks and circle the correct choice.

Example

independent = hours; *dependent* = pay; *As* hours *increase*, pay *increases*/*decreases*.

1. Tanya is trying to coordinate how many buses, b, she needs to transport all the students, s, to the La Brea Tar Pits on a field trip next month, but all the students haven't turned in their permission slips yet.

   independent = _____; *dependent* = _____; *as* _____ *increase(s)*, _____ *increases/decreases*

2. Louisa manages the school cafeteria and she has noticed a relationship between the number of cans of soda, s, that kids consume, and the ability of kids to focus, f, in their afternoon classes.

   independent = _____; *dependent* = _____; *as* _____ *increase(s)*, _____ *increases/decreases*

3. Zach notices that his yield of vegetables has improved since he added more earthworms to his garden. Use v for the number of vegetables he gets each week and w for the number of earthworms in his garden.

   independent = _____; *dependent* = _____; *as* _____ *increase(s)*, _____ *increases/decreases*

4. Dr. Ezra suggests that all of his clients run or walk every day for exercise to improve their general fitness. The variable e is the number of minutes they exercise and d is the number of days they miss from work due to illness.

   independent = _____; *dependent* = _____; *as* _____ *increase(s)*, _____ *increases/decreases*

5. Auggie is getting car insurance quotes before buying his first car. He notices that the insurance quotes are cheaper as the age of the cars increases. Use i for the price of insurance and a for the age of the car.

   independent = _____; *dependent* = _____; *as* _____ *increase(s)*, _____ *increases/decreases*

6. Milo goes on a lot of business trips for his jobs. There is a relationship between the number of T-shirts he packs, t, and the number of days he is on the road, d.

   independent = _____; *dependent* = _____; *as* _____ *increase(s)*, _____ *increases/decreases*

Analyzing Algebraic Relationships

When algebraic relationships contain two variables, the independent variable acts, and the dependent variable *reacts*. Knowing if the dependent variable is increasing or decreasing as the independent variable increases gives you a good picture of how the independent variable influences the dependent variable. Algebraic relationships are frequently used to create tables and graphs that can be used to analyze the relationship between their variables.

Creating Tables

The equation $p = \$12 \times h$ represented your pay, p after h hours of babysitting for \$12/hr. We calculated that if you worked one hour you'd earn \$12 and if you worked two hours you would earn \$24. There are many more combinations of hours and pay that would make the equation $p = \$12 \times h$ true, but let's start by putting the first two pairs of values in a table. When making vertical tables, the independent variable goes in the first column, and the dependent variable goes in the second column. This demonstrates that the value on the left is determining the value on the right. When making a horizontal table, the independent variable goes in the top row, and the dependent variable goes in the bottom row:

Hours h	Pay p
1	\$12
2	\$24

Hours h	1	2
Pay p	\$12	\$24

The equation $p = \$12 \times h$ can be used to create more pairs of solutions to add to the table. Simply chose values for the independent variable, h, and plug them into the equation to generate new values for p. Suppose you plan on working from 6:00 P.M.–10:30 P.M. on Saturday night and then from 10:00 A.M.–6:00 P.M. on Sunday. Since you will be working 4.5 hours on

Saturday, plug 4.5 in for h and solve for p: $p = \$12 \times 4.5 = \$54$. Then plug eight hours into the equation to determine your Sunday pay: $p = \$12 \times 8 = \$96$. Let's add these two new pairs of solutions to the table:

Hours h	Pay p
1	\$12
2	\$24
4.5	\$54
8	\$96

Using Tables to Make Estimates

Tables are great ways to see trends and to make quick estimates. If the next weekend you didn't want to work too hard, but wanted to earn \$75 to buy your mom a nice birthday gift, you could see from the table that you'd need to work somewhere between 4.5 and eight hours. You might make a fast estimate that six hours will get you enough money and when you do $\$12 \times 6 = \$72$, you'll see you were very close!

Example: Tanya realized that there are no buses available for the school field trip, so she is using parent volunteers. For every four students, s, she will need one parent volunteer, v. Since every four students create the need for one more volunteer, Tanya knows that she can use the relationship $v = \frac{1}{4}s$ to determine how many volunteers she'll need. Megen told Tanya there were 35 kids going on the trip, but Becca told her there were 37 kids signed up for the trip. And Tassie thinks that they have 40 confirmed students. How confusing! Make a table representing all these different possibilities and determine how many volunteers Tanya should be looking for.

Solution: We need to plug the number of kids reported by each teacher into the equation for s and then solve for v. When setting up our table, make sure students go in the first column since they will determine how many volunteers are required.

Students s	Volunteers v
35	8.75
37	9.25
40	10

In a previous lesson we reflected on what kinds of numbers made sense for certain variables. You should remember that in certain situations only whole numbers make real-world sense. This is definitely one of those cases since Tanya cannot request 8.75 volunteers! If 35 students are going on the field trip, then nine cars will be needed, since 8.75 rounds up to nine. It might not be as obvious that if 37 students are attending, we need to round 9.25 volunteers up to 10, even though numerically, 9.25 is *closer* to nine. This is because nine cars will only transport 36 kids, and the 37th student will need another car to ride in. So, whether there are 37 students or 40 students, Tanya will need 10 volunteers in either case.

Practice 2

1. Fill in the following table to model the relationship between m and n: $m = n + 4$

n	m
3	
4.5	
6.8	

2. Create a table of values that demonstrates the relationship between w and v. Choose three positive values of v that are less than 10.
$w = 2.5v$

3. Looking at the equation $6 + d = e$, which variable is the independent variable and which is the dependent variable? Make a table to illustrate the relationship between d and e, selecting three values for the independent variable that are between 10 and 20.

4. Kevin has had his fish tank of guppies for many years and he notices that about one guppy dies each month. If his tank currently has 20 guppies and he does not plan on buying any new guppies, write an equation that models f fish he will have after m months. Create a table of values for the relationship between m and f. Which types of values make real-world sense for m and f?

Creating Graphs

Graphs are another way to see clearly see how the two variables relate to each other. Let's return to the babysitting earnings equation $p = \$12 \times h$. In order to graph this relationship, let's look at the table we created:

Hours h	Pay p
1	$12
2	$24
4.5	$54
8	$96

Plot values from tables by creating coordinate pairs where the first column represents the x-coordinates and the second column represents the y-coordinates. We will make the following coordinate pairs from the table: (1,12); (2,24); (4.5,54); (8,96). Noticing that the pay increases by 12 every hour, we can decide to count by 12 on our y-axis so that we can have enough room for all the coordinate pairs:

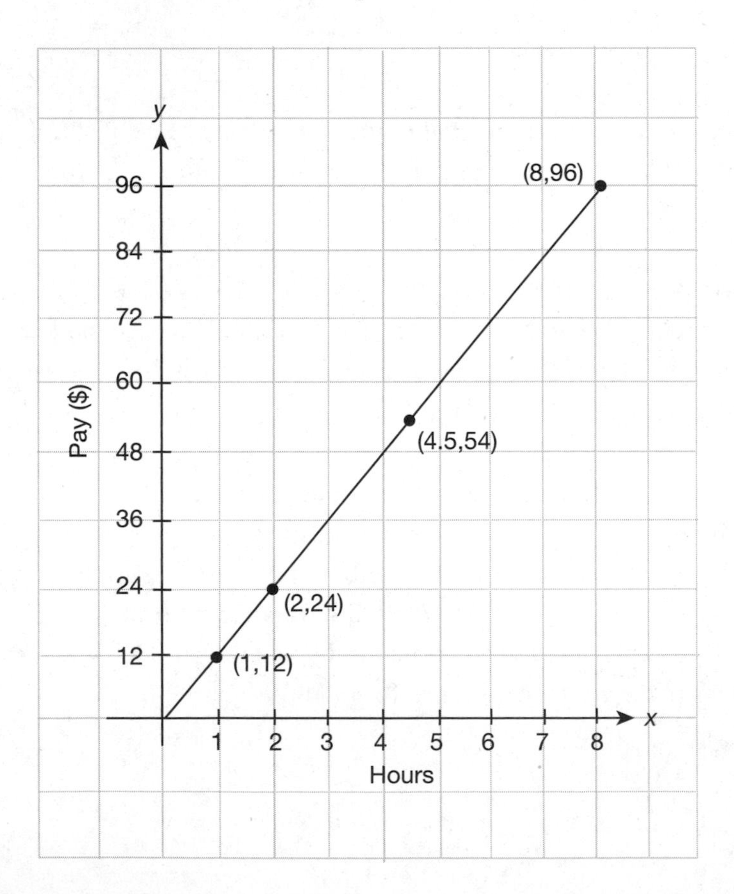

This graph is a useful tool for getting more information about the relationship. We can use it to see how much you'd need to work to earn a specified dollar amount, or we could use it to determine how much money you'd earn after working a set number of hours. We will practice both of these techniques in later lessons.

Practice 3

Use the following information to answer questions 1 through 6.

> *Alton is training hard for a bike race. Every weekend he heads into the San Jacinto mountains for some serious cycling for as many hours as his schedule will allow. Despite the steep hills he climbs, Alton averages 20 miles per hour for h hours each Saturday and covers m miles in a single day.*

1. Which variable is the independent variable and which is the dependent?

2. Write a relationship that expresses m number of miles Alton bikes on a given Saturday after riding for h hours.

3. Use the relationship between m and h to generate a table of values of at least three pairs of data. Make sure you put your dependent and independent variables in the correct columns.

4. Use the table you generated to graph ordered pairs on the coordinate plane.

5. Connect the points graphed in question 4 to generate a line showing the relationship between time and distance. Determine two new ordered pairs using x-coordinates 2.5 and 3.25. What do these coordinate pairs represent?

6. What types of numbers are realistic for h? Why are there limitations on the values for this variable?

Answers

Practice 1

1. *independent = s; dependent = b; as students increase, number of buses increases.*
2. *independent = s; dependent = f; as soda consumed increases, focus decreases.*
3. *independent = w; dependent = v; as number of earthworms increases, number of vegetables increases.*
4. *independent = e; dependent = d; as exercise increases, days of work missed decreases.*
5. *independent = a; dependent = i; as age of car increases, the insurance quote decreases.*
6. *independent = d; dependent = t; as number of days of trip increases, number of T-shirts increases.*

Practice 2

1.

n	*m*
3	7
4.5	8.5
6.8	10.8

2.

v	*w*
0	0
1	2.5
2	5

3. The variable d is the independent variable and e is the dependent variable.

d	*e*
10	16
11	17
20	26

4. $f = 20 - m$. m should be a whole number that is less than 20, since after 20, f would become negative and it doesn't make sense to have a negative number of fish. f will also be a whole number since Kevin can't have a fraction of a fish.

m	*f*
5	15
10	10
20	0

Practice 3

1. Alton is not allowing his mileage to determine his time, but instead he is biking for as many hours as possible each weekend. Therefore, hours, h, is the independent variable and mileage, m, is the dependent variable.

2. $m = 20h$

3.

Hours h	Miles m
1	20
2	40
3	60

4.

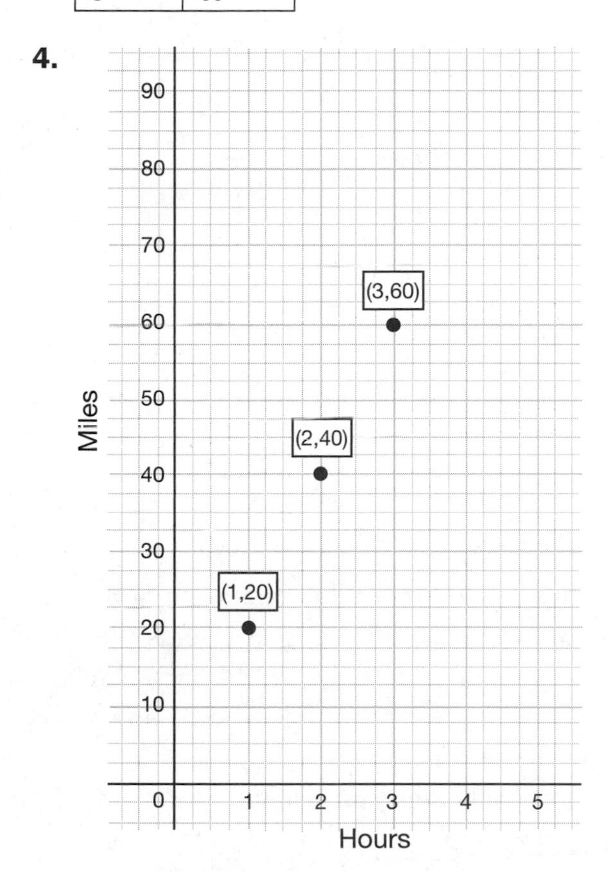

5. Connecting the points gives this line, which shows the two labeled points:

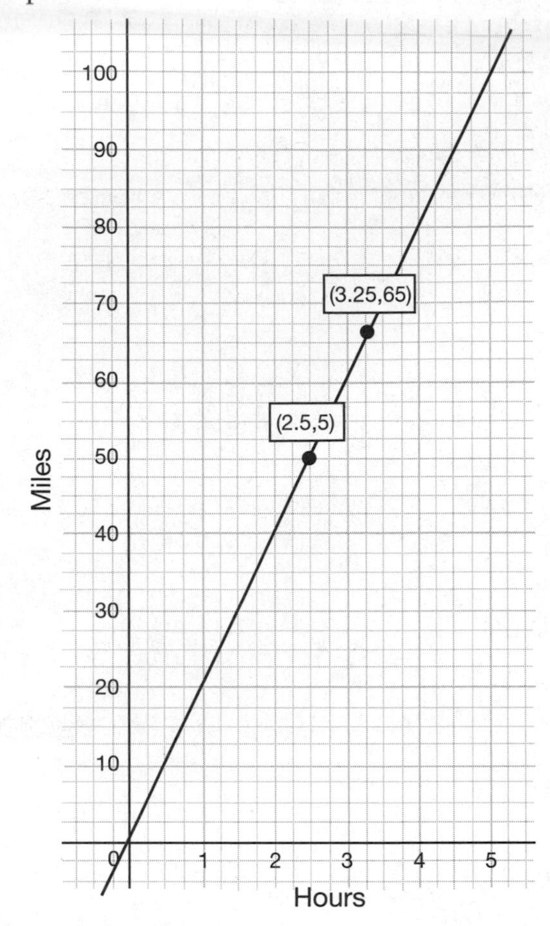

Starting at the x-axis, move out to the x-value of 2.5 and notice that the line has a height of 50 when $x = 2.5$. This represents the point (2.5,50), which means that Alton went 50 miles in 2.5 hours. Similarly, the points (3.25,65) show that Alton went 65 miles in 3.25 hours. (We will practice this more in later lessons.)

6. h can be decimal or fractional values greater than or equal to 0. Since Alton probably wouldn't bike more than 10 hours in a training day, it is reasonable to say that h should be less than 10.

11
Understanding and Graphing Proportional Relationships

STANDARD PREVIEW

In this lesson we will cover **Standards 7.RP.A.1**, **7.RP.A.2.A**, and **7.RP.A.2.D**. You will learn how to determine whether two quantities are proportional by testing them in three different formats: tables, coordinate pairs, and on a coordinate plane. You will also learn which coordinate pairs are the most important points on the graphs of proportional relationships.

Understanding Proportionality

You already know that a ratio is a comparison of two different quantities and in this lesson we will learn how to tell if two ratios are proportional.

Ratios are proportional if they represent the same relationship and are equivalent fractions.

Example: *You ate one slice of pizza in five minutes and your brother ate two slices of pizza in 10 minutes. Is he eating faster than you? Are your eating rates proportional?*
Solution: Your rate is $\frac{1 \text{ slice}}{5 \text{ minutes}}$. Your brother's rate is $\frac{2 \text{ slices}}{10 \text{ minutes}}$, which reduces to $\frac{1 \text{ slice}}{5 \text{ minutes}}$. Therefore, you are both eating at the same rate of one slice every five minutes and the relationship between your eating rates is proportional.

Ratios are proportional if they represent the same relationship and can be reduced to equivalent fractions. If two ratios are not identical when reduced to simplest terms, then they are not equivalent and are not proportional.

Finding Proportional Relationships in Tables

In order to check to see if two quantities represented in a table have a proportional relationship, we need to see if a multiplication relationship exists between the pairs of data. In an earlier lesson we used multiplication relationships to solve for missing values, but now we will test given values to see if a multiplication relationship exists. The following table models the number of tablespoons of protein powder needed to make a smoothie with water.

Water (*c*)	Protein Powder (*T*)
1	3
2	6
5	15

The first row shows there is a ratio of 1:3. If we double the water and double the protein powder, we arrive at the second ratio, 2:6, so a multiplication relationship exists between the ordered pairs (1,3) and (2,6). If we multiply the first ratio 1:3 by five we end up with the third ratio, 5:15, so a multiplication relationship exists there as well:

$$\frac{1 \text{ c water}}{3 \text{ T powder}}\left[\frac{\times 2}{\times 2}\right] = \frac{2 \text{ c water}}{6 \text{ T powder}}$$

$$\frac{1 \text{ c water}}{3 \text{ T powder}}\left[\frac{\times 5}{\times 5}\right] = \frac{5 \text{ c water}}{15 \text{ T powder}}$$

When a multiplication relationship exists between two fractions, the two fractions are equivalent. Since all the ordered pairs in the table are equivalent to each other, we would say *the amount of protein powder is proportional to the water used.*

Sometimes the values in a table will not be so easy to verify using multiplication. Consider the following table of values, which shows Liv's walking routine over three separate days.

Minutes	5	7	11
Meters	400	560	880

There are no easy multiplication relationships between these points, but that doesn't mean they are not proportional. In order to work with a table of values like this, it is necessary to find the unit rate of each of the ordered pairs in the table. Remember that **unit rate** is a rate that compares two different units of measurement that has a denominator of one. Since the independent variable is determining the value of the dependent variable, you want your statement to be in terms of one unit of the independent variable. Therefore, when finding the unit rate of ordered pairs, put the dependent variable in the numerator and the independent variable in the denominator. Then divide both parts of the rate by the denominator to create the unit rate with 1 in the denominator:

Find the unit rate for (5,400): $\frac{400 \text{ meters}}{5 \text{ minutes}}\left[\frac{\div 5}{\div 5}\right] = \frac{80 \text{ meters}}{1 \text{ minute}}$

Find the unit rate for (7,560): $\frac{560 \text{ meters}}{7 \text{ minutes}}\left[\frac{\div 7}{\div 7}\right] = \frac{80 \text{ meters}}{1 \text{ minute}}$

Find the unit rate for (11,880): $\frac{880 \text{ meters}}{11 \text{ minutes}}\left[\frac{\div 11}{\div 11}\right] = \frac{80 \text{ meters}}{1 \text{ minute}}$

Since each of these rates reduced to the rate of $\frac{80 \text{ meters}}{1 \text{ minute}}$ or 80 meters per minute, Liv's distance covered is proportional to the time she spends walking. Although she is walking different distances each day, she is walking at the same rate, or speed.

In order to find the unit rate of ordered pairs in proportional relationships, divide the dependent variable, y, by the independent variable, x.

$$\text{Unit Rate} = \frac{\text{dependent variable } (y)}{\text{independent variable } (x)}$$

Finding Proportional Relationships between Coordinate Pairs

Dylan wants to make four liters of lemonade by adding three cups lemonade mix to four liters of water. His sister Luna insists they make six liters of lemonade and wants to use 4.5 cups of lemonade mix. How can we tell if (4,3) and (6,4.5) are proportional and will therefore make equally delicious lemonade? Find the unit rate as we did previously for each pair by dividing the dependent variable by the independent variable. Since the amount of water used is *determining* the number of cups of lemonade mix used, the water is the independent variable and the lemonade is the dependent variable.

Find the unit rate for (4,3): $\frac{3 \text{ cups}}{4 \text{ liters}}\left[\frac{\div 4}{\div 4}\right] = \frac{\frac{3}{4}\text{ cup}}{1 \text{ liter}}$

Find the unit rate for (6,4.5): $\frac{4.5 \text{ cups}}{6 \text{ liters}}\left[\frac{\div 6}{\div 6}\right] = \frac{\frac{3}{4}\text{ cup*}}{1 \text{ liter}}$

Since both of these recipes would be using $\frac{3}{4}$ cup of lemonade mix for every liter of water, the lemonade mix is proportional to the water in both of these ordered pairs.

*Here's a reminder of how to divide 4.5 by six using improper fractions:

$$4\frac{1}{2} \div \frac{6}{1} = \frac{9}{2} \div \frac{6}{1} = \frac{9}{2} \times \frac{1}{6} = \frac{9}{12} = \frac{3}{4}$$

ERROR ALERT! It's a common mistake to try to find the unit rate by dividing the x-variable by the y-variable, but make sure you always put the independent variable in the denominator!

$$\textbf{Unit Rate of Proportional Relationships} = \frac{\text{dependent variable } (y)}{\text{independent variable } (x)}$$

Practice 1

Use the following information for questions 1 and 2:

Miana drove 224 miles in four hours on Monday. Then on Tuesday she drove for 6.5 hours and covered 364 miles.

1. Create two ordered pairs representing Miana's travels, paying special attention to which quantity should be your independent variable and which should be your dependent.

2. Use your ordered pairs to determine if Miana's time and distance traveled was proportional from Monday to Tuesday.

3. Does the following table show a proportional relationship between the number of babies in day care and the number of diapers changed?

# Babies	2	4	6	8
# Diapers	5	10	16	20

4. What could you change in the table to make all of the ordered pairs proportional?

Use the following information for questions 5 through 7:

The ordered pairs $(10,\frac{1}{2})$ *and* $(25,1.25)$ *represent the duration of time Jayco walked and the distance he covered.*

5. Use what you know about independent and dependent variables to determine which of the numbers in the ordered pair represents a unit of time and which number represents a unit of distance. Make a hypothesis for what each coordinate could be representing in terms of the units of distance and time (Using units like minutes, feet, miles, or hours, what units make the most sense for these ordered pairs?)

6. Determine if the two points represent a proportional relationship between distance and time.

7. Determine a unit rate for Jayco and explain what it means.

Proportional Relationships on Graphs

A graph can be used to determine if a collection of ordered pairs are proportional.

> **Example:** Kaila and Sanaa are doing a science experiment to find out if shadow length is proportional to height. Using their siblings as volunteers, the girls measure each person's height and record it in the table. Then, at precisely 4:00 P.M., they measure the length of each person's shadow and put all the data in the table shown below. (Notice that shadow length *depends* on the height, so height is the independent variable in the first column.)

	Height (cm)	Shadow Length (cm)
Kayla	183	244
Angelina	162	216
Malu	93	124
Jack	99	132

To create a graph, label the x-axis as *Height*, and the y-axis as *Shadow Length*, and plot the four different ordered pairs on the coordinate plane:

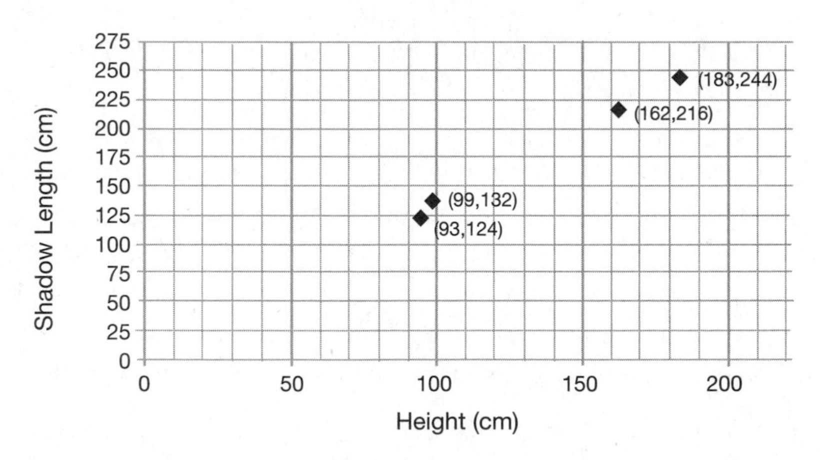

Points on a coordinate plane have proportional relationships if they fall on the same straight line, and if that line goes through the origin (0,0). Since these four points make a straight line, Kaila and Sanaa have determined that shadow length is proportional to height.

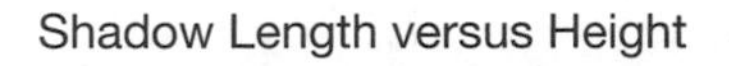

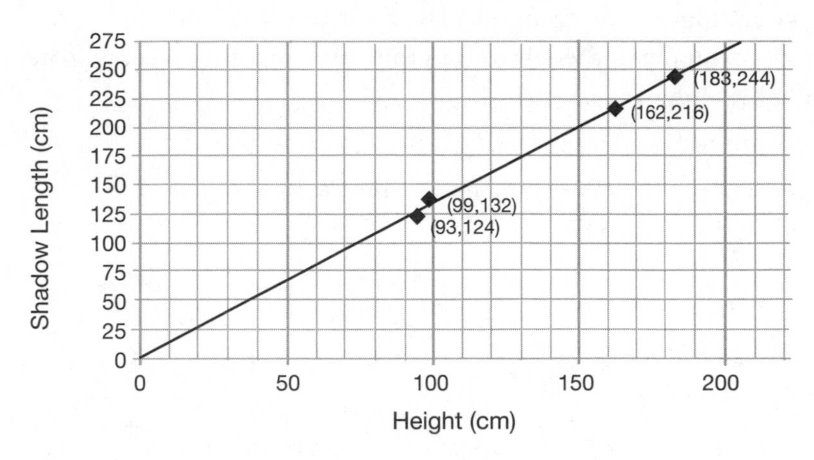

Two or more quantities have a proportional relationship if the ordered pairs representing them are on the same straight line, which goes through the origin (0,0).

Interpreting Points on Graphs

Graphs are powerful tools for helping us make comparisons and predictions when we have limited information about a situation. Only four pairs of height and shadow measurements were used to create the line in the Shadow Length vs. Height graph, but we can use this graph to get much more information. Because shadow length and height have a proportional relationship, any point on the line tells us how long the shadow would be for a person of a certain height. For instance, since the line passes through (150,200), we know that a person who is 150 cm tall will have a shadow that's 200 cm long.

Any (x,y) point on the graph of a proportional relationship represents a point where the independent variable (x) has directly determined the value of its corresponding dependent variable (y).

Using Graphs to Make Predictions

Let's use the previous line to make predictions about the shadow lengths of people of different heights. If you want to know how lon g the shadow would be of a person who is 110 centimeters tall, find 110 on the x-axis, move your finger straight up until you hit the line, and then move your finger horizontally to the left to find the corresponding y-coordinate. The graph below shows this process:

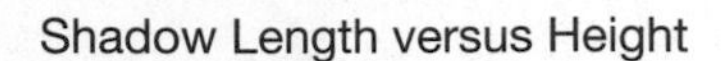

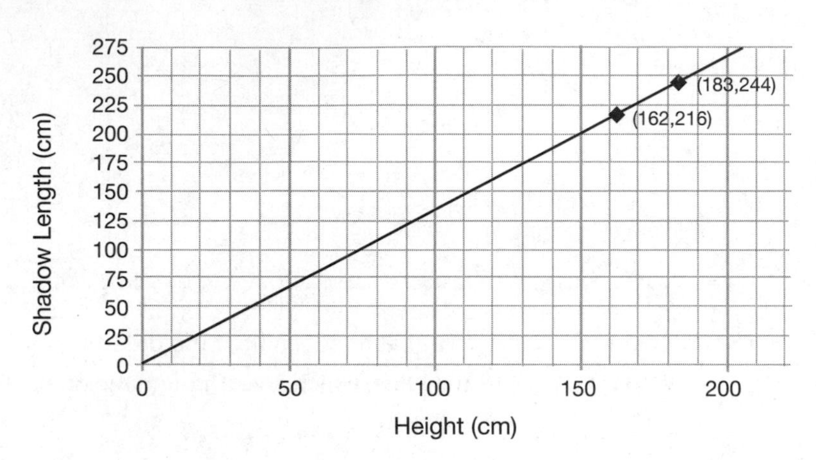

A person 110 centimeters tall will have a shadow about 150 centimeters long. How could you reverse this process to determine that someone with a shadow of 75 centimeters is around 55 cm tall? Simply find a shadow length of 75 cm along the y-axis, move your finger to the right, and see where it meets the appropriate height on the x-axis (55 cm).

The Importance of the Origin

Now notice that the shadow length line passes through the origin, (0,0). This is a feature of proportional relationships: they always include the point (0,0). If an object were zero centimeters tall, it will have a shadow zero centimeters long. The same is true for any proportional relationship. If you buy zero gallons of gas, you pay $0.00. If you bake zero cookies, you will need zero cups of sugar. *It is important to remember that in order for a relationship to be proportional, its graph must go through the point* (0,0).

The Importance of the Point (1,*r*)

The following graph represents how far Paige has driven over the course of time in hours. It goes through the origin, and all the data is on the line, so we know this is a proportional relationship. Noticing that each finer line on the vertical axis is counting by 20 miles, use the graph to fill in the corresponding distances for the following hours driven: (5,____), (3,____), and (1,____).

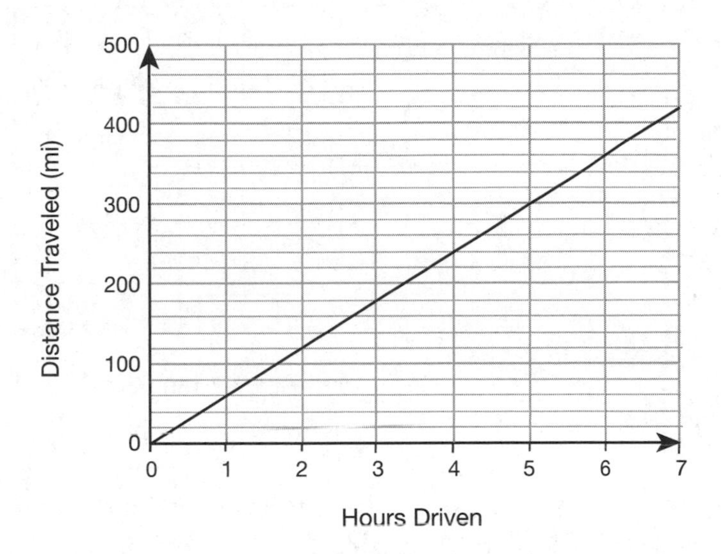

You should have concluded that in five hours Paige drove 300 miles, in three hours she drove 180 miles, and in one hour she drove 60 miles. Using that last coordinate pair, we can determine that if Paige drove 60 miles in one hour, her *unit rate, or speed, was 60 miles per hour.* Using a graph to determine the unit rate of a proportional relationship is easy: just determine the value of y when x is one. Your independent variable is always x, so the value of the dependent variable, y, when x is one, will always be your unit rate. This point is referred to a $(1,r)$, where r equals the unit rate.

The graphs of all proportional relationships must go through the point (0,0). All proportional relationships will also contain the point (1,*r*) where *r* is the unit rate.

Sometimes the data has a scale that is too large to get an exact reading for the value of y when $x = 1$. This is certainly the case with the Shadow Length vs. Height graph in the earlier example. Looking at the graph below, it would not be accurate to use the graph to find the coordinate of the point $(1,y)$.

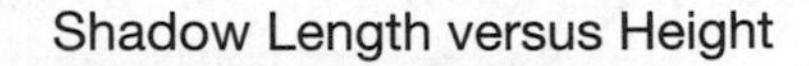

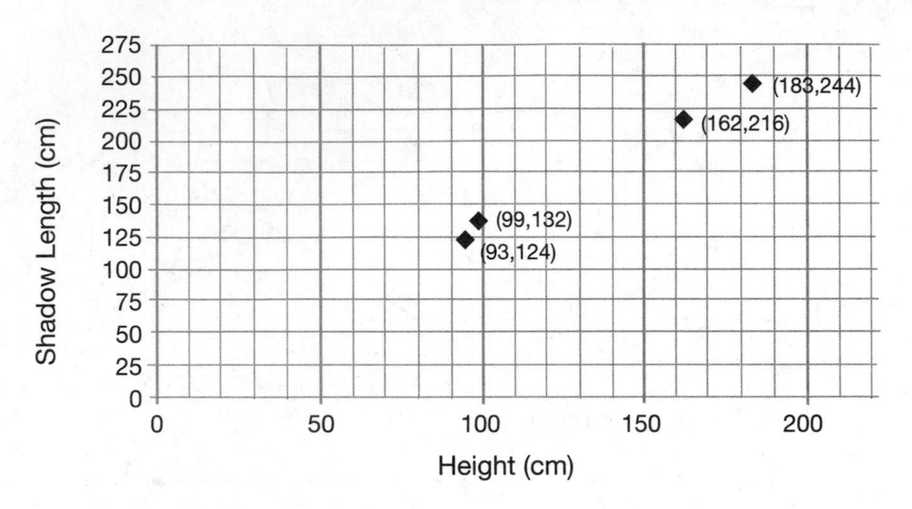

However, since you know the point (162,216), you can represent that as the ratio $\frac{216}{162}$ (putting the dependent variable in the numerator) and then divide both parts of this ratio by 162 to reduce it to the unit rate:

$$\frac{216 \text{ cm shadow}}{162 \text{ cm height}}\left(\frac{\div 162}{\div 162}\right) = \frac{\frac{4}{3}\text{ cm shadow}}{1 \text{ cm height}}$$

Therefore, we know that the point $(1,\frac{4}{3})$ is on the graph. This point can be interpreted to mean *for every 1 cm of height, a person has $\frac{4}{3}$ cm of shadow.* (Since the earth's relationship to the sun changes every day, this height-shadow relationship is not a constant relationship at all times, but just at 4:00 P.M. on the day that Kaila and Sanaa did their measurements.)

Practice 2

Use the following table to complete questions 1 through 3.

Bags of Potatoes	Weight in Pounds
2	3
5	7.5
8	12
7	
	5

1. Plot the first three given ordered pairs on the coordinate plane and determine if they have a proportional relationship.

2. Use the point $(1,r)$ to find the unit rate.

3. Use the line to find out the weight of seven bags. Then determine approximately how many bags of potatoes it will take to weigh five pounds. Fill these values in the preceding table.

Use the information and graph to complete questions 4 through 6:

Over the summer Nick takes delivery orders over the phone at the local pizza shop. The following graph represents the relationship between his pay and hours:

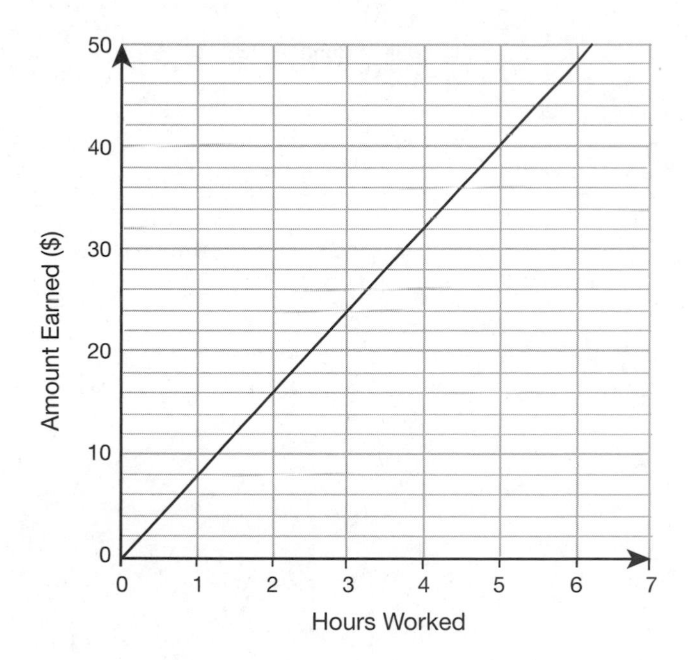

4. If Nick works three hours on Saturday how much will he earn?

5. Nicks wants to make $45 on Saturday so he can take his friend to a concert for his birthday that night. How many hours does he need to work?

6. Use the point $(1,r)$ to find Nick's earnings per hour.

The following graph shows how much Melinda pays for bananas at Trader Jeffs. Use this graph to complete questions 7 and 8:

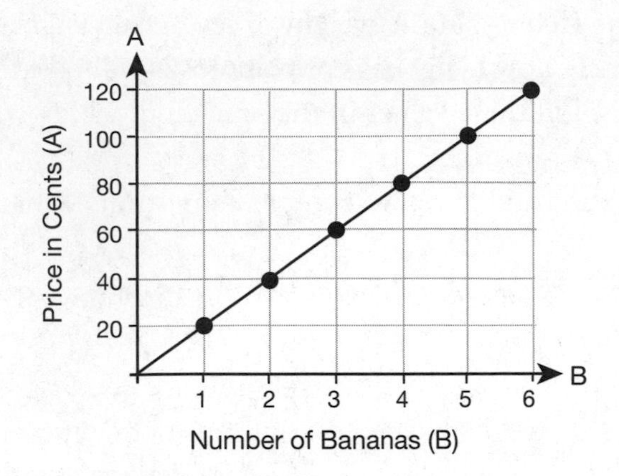

7. What does the point (6,120) tell you?

8. Which point can be used to determine the unit price of bananas at Trader Jeffs? What do they charge per banana?

Answers

Practice 1

1. (4,224) and (6.5,364). The time she drives determines the distance she covers.
2. To determine if Miana's distance was proportional, write the ratios as fractions and see if they are equivalent. Since the second day will have 6.5 in the denominator, making it hard to reduce, it is easiest with this data to find the unit rate of each pair:

   Monday Unit Rate = $\frac{\text{dependent variable } (y)}{\text{independent variable } (x)} = \frac{224}{4} = \frac{56}{1}$, 56 miles/hour

   Tuesday Unit Rate $\frac{\text{dependent variable } (y)}{\text{independent variable } (x)} = \frac{364}{6.5} = \frac{56}{1}$, 56 miles/hour

   Yes, Miana's time and distance relationship between Monday and Tuesday is proportional.
3. Although the second and fourth columns are equivalent to the initial ratio of $\frac{2}{5}$, the third column's ratio of $\frac{6}{16}$ is not equivalent to $\frac{2}{5}$

# Babies	2	4	6	8
# Diapers	5	10	16	20

4. If the third coordinate pair was six babies and 15 diapers, then the relationship between babies and diapers would be proportional throughout the table.
5. The first coordinate is could be the time in minutes and the second coordinate is the distance in miles.
6. Check if the ratios are equivalent to determine if the coordinate pairs are proportional: The coordinate pair $(10,\frac{1}{2})$ can be written as $\frac{10}{.5}$ and then using a multiplication relationship of two and then of 1.25, creates the equivalent fraction $\frac{25}{1.25}$, so these pairs are proportional.

$$\frac{10}{0.5}\left(\frac{\times 2}{\times 2}\right) = \frac{20}{1}\left(\frac{\times 1.25}{\times 1.25}\right) = \frac{25}{1.25}$$

7. Unit Rate $\frac{\text{dependent variable } (y)}{\text{independent variable } (x)} = \frac{0.5 \text{ mile}}{10 \text{ min}} = \frac{0.05 \text{ mile}}{1 \text{ min}}$, 0.05 miles/minute. Jayco is walking at a speed of 0.05 miles per minute. Since *miles per minute* is not a standard measurement used, we could change this to *miles per hour* by multiplying both of the parts of the ratio by 60:

   $\frac{0.05 \text{ mile}}{1 \text{ min}}\left(\frac{\times 60}{\times 60}\right) = \frac{3 \text{ miles}}{60 \text{ min}}$ = 3 miles/hour

Practice 2

1. Since these three points all fall on the same line and the line goes through (0,0), we know that these points are proportional:

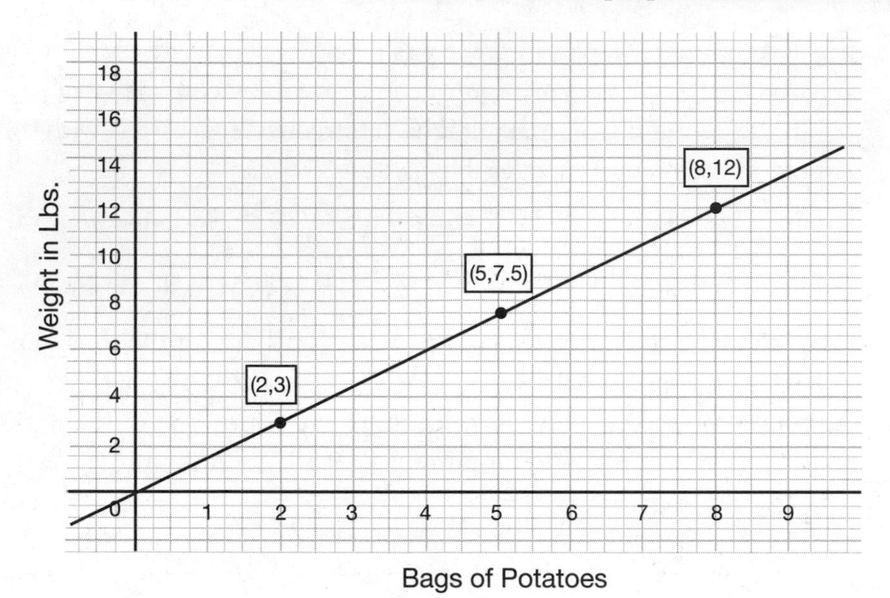

2. When $x = 1$, $y = 1.5$ so the unit rate is 1.5 pounds per bag.
3. The weight of seven bags is 10.5 lbs. It would take a little less than 3.5 bags to weigh five pounds.

Bags of Potatoes	Weight in Pounds
2	3
5	7.5
8	12
7	10.5
Estimate 3.4	5

4. If Nick works three hours on Saturday he will earn $24.
5. Nick needs to work between 5.5 and six hours to earn $45.

6. The point (1,8) shows Nick's earnings at $8 per hour.

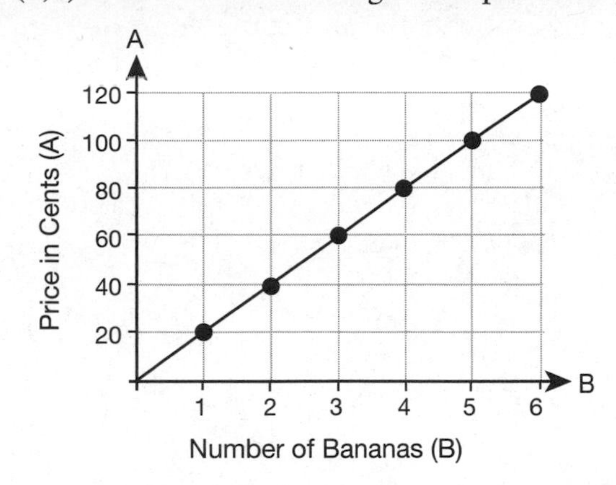

7. The point (6,120) shows that six bananas cost 120 cents or $1.20.

8. The point (1,20) shows that the unit cost is $0.20 per banana.

12 Writing Proportional Relationship Equations

STANDARD PREVIEW

In this lesson we will cover **Standards 7.RP.A.2.B** and **7.RP.A.2.C**. You will learn how to represent proportional relationships in algebraic equations. You will learn how to determine the *constant of proportionality* in tables, graphs, equations, and word problems, and how it is used to construct equations.

The Constant of Proportionality

The **constant of proportionality** is a fancy name for something you are already very familiar with: unit rate. The unit rate is expressed in "per unit" language, like *60 miles per hour*. The constant of proportionality is not expressed with *per unit* language and is instead expressed as a single number that is then used in equations. If the unit rate is *60 miles per hour*, the constant of proportionality is 60. The general form for proportional equations is:

[Dependent variable] = [**Constant of Proportionality**]
× [Independent variable]

When the constant of proportionality is p, the proportional equation is $y = px$.

The constant of proportionality is the same as the unit rate, but it is expressed as a single number. It is used to write proportional equations in the form $y = px$, where x = independent variable, y = dependent variable, and p = constant of proportionality.

p will always be equal to $\frac{y}{x}$ in proportional relationships.

The Constant of Proportionality in Written Descriptions

Since a lot of the math you will face in the real world will come from written or verbal descriptions (as opposed to equations just falling in your lap), it's important to be able to translate given information into mathematical equations. The trickiest thing can sometimes be identifying which unit is the independent variable.

Example: *Kyle paid $8 for 2.5 pounds of organic guava fruit. Find the constant of proportionality and write a proportional equation representing the relationship between the weight and cost of guava.*
Solution: Here, the pounds of guava purchased will determine the total price, so make *pounds* your independent variable in the bottom of your unit rate ratio. First calculate the constant of proportionality by dividing the dependent variable by the independent variable and reducing it so that we just have 1 lb. in the denominator:

- $\frac{\$8.00}{2.5 \text{ lbs.}}\left(\frac{\div 2.5}{\div 2.5}\right) = \frac{\$3.20}{1 \text{ lb.}}$
- The unit rate is $3.20/pound so the constant of proportionality, p, is $3.20.
- Using the equation $y = px$, substitute $3.20 in for p: $y = \$3.20x$

In order to give the equation more context, you can instead use g to represent the weight of guava and C to represent the total cost: $C = \$3.20g$.

Practice 1

1. Compare and contrast the constant of proportionality to unit rate.

2. Write an algebraic equation using the three terms *independent variable*, *dependent variable*, and *constant of proportionality*.

3. Max took a typing test and he typed 215 words correctly in five minutes. Find his unit rate and the constant of proportionality.

4. Use the information from question 3 to write an equation. Choose variables that relate to words and minutes and define what each variable stands for.

5. The Shri Digambar Jain temple in New Delhi, India, houses a bird hospital that rescues and cares for pigeons that have been injured by vehicles or bikes in the city. For every two birds that are brought to the hospital, the hospital needs to use an extra three ounces of feed each day. Write an equation to represent this relationship using b for birds and f for feed.

6. Referring to the equation you made in question 5, what types of values make sense for b?

7. Use the equation you made in question 5 to determine how many extra ounces of feed the Jain temple will need if they receive 13 new birds on Wednesday.

The Constant of Proportionality in Tables

If the values in a table have proportional relationships (they all reduce to the same ratio), then you can determine the constant of proportionality and write an algebraic equation to model that relationship.

Example: *The following table shows the prices for bagels at Brooklyn Bagels. Find the constant of proportionality and write an equation that models the proportional relationship.*

Bagels	12	24	36	48	60
Price	$6	$12	$18	$24	$30

Solution:

- In this case, $\frac{\$6.00}{12 \text{ bagels}}\left(\frac{\div 12}{\div 12}\right) = \frac{\$0.50}{1 \text{ bagel}}$, so the unit price is $0.50 per bagel and the constant of proportionality is 0.50.
- We can use the constant of proportionality to write the following algebraic equation:
  $P = 0.50d$, where d = number of bagels and P = total price
- This equation tells us that we can multiply the number of bagels by $0.50 to find out the total price.

Notice in the last equation that we explained what each of the variables represents:

$P = \$0.50d$, where d = number of bagels and P = total price

This is called defining your variables and it is a critical step when writing algebraic equations.

We started you out with a tricky example because the unit rate was not apparent in the table of the price of bagels. Remember that when a table has a value of 1 for the independent variable, the constant of proportionality will just be the corresponding y value.

Example: The following table shows the maximum number of pounds a sport fishing boat can bring to shore each day depending on how many paying clients they take out fishing. Test to see if it is proportional and if it is, write a relationship representing how the pounds of fish relates to the number of clients.

Number of Clients	1	2	5	10
Pounds of Fish	30	60	150	300

Solution: First we notice that by using the point (1,30) we can multiply both values in this ratio by two, five, or 10, respectively, to get the other three points. Since there is a multiplication relationship we know these ordered pairs are proportional.

$$\frac{1}{30}\left(\frac{\times 2}{\times 2}\right) = \frac{2}{60}$$
$$\frac{1}{30}\left(\frac{\times 5}{\times 5}\right) = \frac{5}{150}$$
$$\frac{1}{30}\left(\frac{\times 10}{\times 10}\right) = \frac{10}{300}$$

Next, we look at the point (1,30) to see the constant of proportionality. Since one client on the boat can bring in 30 pounds of fish, the constant of proportionality is 30. Multiply 30 pounds by the number of clients to determine how much fish can be brought to shore:

Letting c = number of clients and F = pounds of fish, the equation is $F = 30c$. (Notice that we defined our variables, c and F. Prepare yourself to get in this habit!)

The Constant of Proportionality in Graphs

We learned in the last lesson that the value r in the point $(1,r)$ is always the unit rate of a proportional relationship. Since we now know that the constant of proportionality is just the unit rate expressed as a single number, we can use this same point to identify the constant of proportionality. Then it's as easy as $y = px$ to write our equation modeling the graph!

Example: Let's take another look at the following graph, which represents how far Paige has driven over the course of several hours. Write an equation that models this graph.

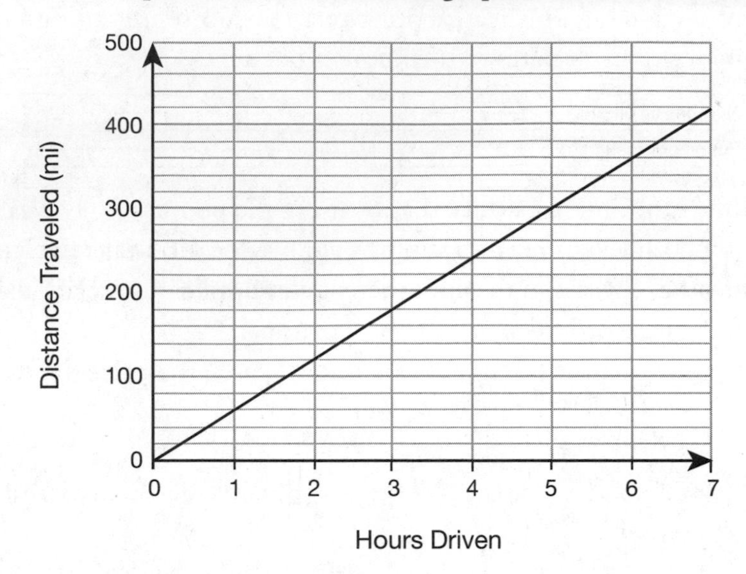

Solution: Looking at this graph carefully it is easy to distinguish the point (1,60) is on the graph. Therefore, the constant of proportionality is 60 and for every hour Paige drives, she will go an additional 60 miles. This graph can be represented as $D = 60t$ where t = time in hours and D = distance in miles.

Large Scale Graphs

Help! The scale on my coordinate plane is too spread out to accurately read the y value when x = 1. If this is the case, don't panic. Remember that you can always find the constant of proportionality in proportional relationships by dividing *any y*-variable by its *corresponding x*-variable. Therefore, a line containing the point (6,12) will have a constant of proportionality of 0.50 by performing: $\frac{\$6.00}{12 \text{ bagels}} = \$0.50$.

Practice 2

1. What is the constant of proportionality in the following table?

Dogs	1	2	3
Cans of Food	20	40	60

2. Sandy's Treasures sells used books in different sized bundles at a great discount. According to the table below, what is the price per book she charges?

Number of Books in Bundle	Price in Dollars
3	$6.75
5	$11.25
12	$27

3. Use your answer to question 2 to write an equation representing the relationship between number of books and price.

4. Use your equation in questions 3 to determine the price Sandy should charge you if you buy seven books.

Use the following information and graph to answer questions 5 and 6.

A nonprofit in Ghana buys school supplies so that disadvantaged children can attend school. The following graph shows the cost of this program per month, depending on how many students are in their program any given month:

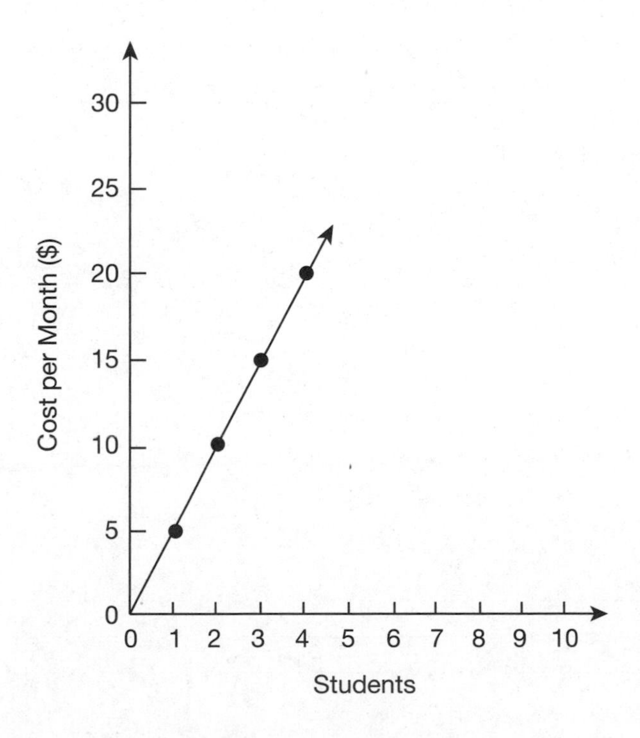

5. What is the constant of proportionality in this graph? How does the constant of proportionality relate to the number of students and cost per month?

6. Write an equation that represents the relationship between number of students and cost per month.

7. The following graph shows how much toilet paper Camp Adams needs to have on hand each week, depending on the number of campers they have signed up. Find the constant of proportionality and explain what that means about their toilet paper needs.

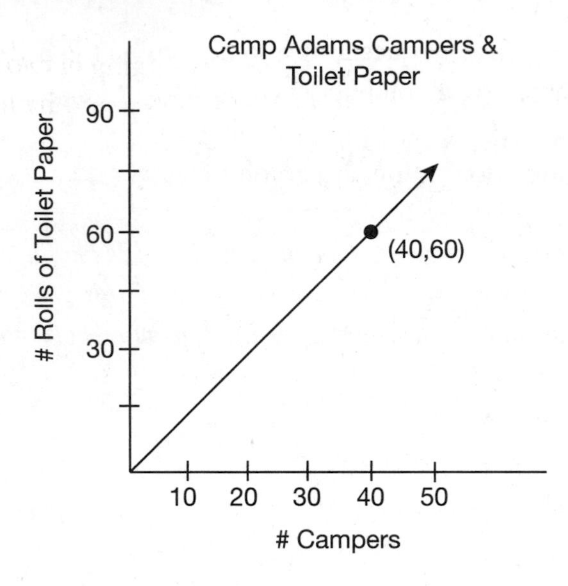

Answers

Practice 1

1. Unit rate is a ratio that compares one type of measurement to a single unit of another type of measurement. It is often expressed in *per unit* language. The constant of proportionality is the same thing as unit rate except it is presented as a single number rather than as a comparison of the two different types of measurements.
2. [Dependent variable] = [Constant of Proportionality] × [Independent variable]
3. Unit rate = 43 words per minute. Constant of proportionality = 43.
4. $w = 43m$, where m = number of words per minute and w = number of words typed.
5. $f = (\frac{3}{2})b$ (This answer can be checked by plugging in two for b and seeing that for two birds, three ounces of food would be needed, which agrees with the verbal description.)
6. b can be any whole number greater than or equal to zero.
7. $f = (\frac{3}{2})b$, for $b = 13$, becomes $f = (\frac{3}{2})(13) = \frac{39}{2}$ or 19.5 ounces of feed.

Practice 2

1. The constant of proportionality is 20.
2. $\frac{6.75}{3} = \frac{2.25}{1}$ so Sandy charges $2.25 per book.
3. $C = 2.25b$
4. $C = 2.25b$, for $b = 7$, shows a cost of $15.75 for seven books
5. The constant of proportionality is 5. For every student in the program, the cost goes up by $5.
6. $C = 5s$ where s = Number of students and C = Cost of program
7. This relationship is proportional since the line goes through (0,0). Divide the dependent variable by the independent variable to find the constant of proportionality: $\frac{60}{40}\left(\frac{\div 40}{\div 40}\right) = \frac{\frac{3}{2}}{1}$. Camp Adams must have 1.5 rolls of toilet paper on hand each week for every student enrolled in their camp.

13

Solving Percents with Proportions

STANDARD PREVIEW

In this lesson we will cover Standard 7.RP.A.3. You will learn how to use proportional relationships to solve problems with percents within the real-world context of tax, tips, commissions, fees, simple interest, and percent increase and decrease.

Setting Up Percentage as Proportions

In Lesson 3 we used rate tables to work with percents, and in this lesson we are going to use proportional relationships to solve more advanced percent problems. A percentage is used to describe a quantity of something; it is a part-to-whole ratio where the percentage is the part and the whole is 100. So if 30% of a shipment of lightbulbs arrived broken, that means that 30 out of every 100 lightbulbs were broken.

But how many bulbs does 30% represent if our shipment was just 60 bulbs? Because percent is a ratio, we can use a proportional relationship to create an equivalent ratio with a "whole" of 60 in the denominator, Then we can use the multiplication relationship to solve for our numerator to determine how many bulbs, out of 60, arrived broken. The following general equation can be used to solve most percentage questions:

$$\frac{\%}{100} = \frac{part}{whole}$$

Example: *EJ ordered 60 specialized lightbulbs for a high-end event she is hosting. The shipping company just told her that the delivery truck had some loading issues and they expect that 30% of her order will arrive broken. How many bulbs will arrive broken?*

Solution:

Step 1: Write the formula.

Start with the general equation for percentage problems:

$$\frac{\%}{100} = \frac{part}{whole}$$

Step 2: Plug in values.

Since we are dealing with 30%, 30 will go over the 100. Next, since 60 was the total number of bulbs that arrived, and only *part* of them were broken, 60 will go in for our *whole* and we will solve for *part*:

$$\frac{30}{100} = \frac{part}{60}$$

Step 3: Find the multiplication factor.

Next, we need to determine how to find an equivalent fraction by looking for a multiplication relationship. Since we don't see one at first, we'll reduce our percentage ratio to simplest form to see if that helps:

$$\frac{30}{100}\left[\frac{\div 10}{\div 10}\right] = \frac{3}{10}, \text{ so } \frac{3}{10} = \frac{part}{60}$$

Now we can see that $10 \times 6 = 60$, so we know to use a multiplication factor of 6.

Step 4: Create the equivalent fraction and answer the question:

Create an equivalent fraction by multiplying by $\frac{6}{6}$:

$$\frac{3}{10}\left[\frac{\times 6}{\times 6}\right] = \frac{18}{60}$$

The new ratio, $\frac{18}{60}$, tells EJ that she should anticipate that 18 of her 60 lightbulbs will arrive broken.

Tax as a Percentage

You will encounter parentages the most in your everyday life with money. When you buy something in most states, there will be an extra charge of sales tax, which is money collected by the state government to fund services. The sales tax is always a percentage of the purchase you are making and is never a flat fee. A 6% tax means you will pay an extra six cents for every 100 cents of an original price. So for every dollar something costs, there will be an additional six cents charged in tax. We can use proportions to solve for sales tax using the same first steps illustrated previously. Sometimes there will be one additional step at the end.

Example: Olivia wants to buy a soy wax candle for $30, but she's not sure if the $32 she has in her wallet will be enough to cover the price with tax. Find the total price of the candle and let Olivia know if she can purchase it today.

Solution:

Step 1: Write the formula.

Start with the general equation for percentage problems:

$$\frac{\%}{100} = \frac{part}{whole}$$

Step 2: Plug in values.

Since tax is 8%, put 8 over the 100. Since tax is a percentage of the purchase price, the $30 price of the candle represents the *whole* and the tax will be the part:

$$\frac{8}{100} = \frac{part}{30}$$

Step 3: Find the multiplication factor.

Since we don't see a multiplication relationship at first, we'll reduce our percentage ratio to see if that helps.

$$\frac{8}{100}\left[\frac{\div 4}{\div 4}\right] = \frac{2}{25}, \text{ so } \frac{2}{25} = \frac{part}{30}$$

There is no easy way to multiply 25 by something to get 30, so let's approach this a different way. If we instead change the denominator of $\frac{8}{100}$ to be a 10, then we can use three as the multiplication factor. Remember that dividing a number by 10 is as easy as moving the decimal place once to the left:

$$\frac{8}{100}\left[\frac{\div 10}{\div 10}\right] = \frac{0.8}{10}, \text{ so } \frac{0.8}{10} = \frac{part}{30}$$

Now we can easily see that $10 \times 3 = 30$, so we know the multiplication factor is three.

Step 4: Create the equivalent fraction and answer the question:

Create an equivalent fraction by multiplying by $\frac{3}{3}$:

$$\frac{0.8}{10}\left[\frac{\times 3}{\times 3}\right] = \frac{2.4}{30}$$

The equivalent ratio, $\frac{2.4}{30}$, tells Olivia that 2.4 or $2.40 will be the tax on her candle.

Step 5: Add the sales tax to the original price:
Since the candle costs $30 and the tax is $2.40, we add the two together to find the total price of $32.40. So unless Olivia can find forty cents in the bottom of her backpack, she needs to say goodbye to the candle for today.

ERROR ALERT! It is a common error for students to think that a 6% tax means adding six cents onto the pretax cost, rather than paying six cents tax for *every* dollar that an item costs. Other times, students will solve for the tax and forget to add it to the original price and instead state that the total price is actually *lower* than the pretax price. Do not fall for either of these errors in thinking!

When working with tax, read the question carefully to see if you are solving for *just the tax* of an item or for the *total cost* of the item with tax. If you are finding the total cost, you will need to add the tax calculated to the original price:

Total Cost = [Original Price in Dollars] + [Sales Tax in Dollars]

Finding the Original Price When Given the Tax

Sometimes with percent questions you will know the tax, but not the original price. Look at how to approach this in the following example:

Example: *Jake's friend Tyler exclaimed with frustration that the tax on his new car was $855 at the DMV. Jake knows that tax is 9.5% and is curious how much Tyler paid for his car.*
Solution:

Step 1: Write the formula.
Start with the general equation for percentage problems:

$$\frac{\%}{100} = \frac{part}{whole}$$

Step 2: Plug in values.
Since tax is 9.5%, we will put 9.5 over the 100. The $855 of tax is the *part* in this question so we'll be solving for the whole:

$$\frac{9.5}{100} = \frac{\$855}{whole}$$

Step 3: Find the multiplication factor.
Find the multiplication relationship by dividing 855 by 9.5: $855 \div 9.5 = 90$. Use 90 as the multiplication factor to create an equivalent fraction.
Step 4: Create the equivalent fraction and answer the question:
Multiplying the numerator and denominator of the percent ratio both by 90 gives us the equivalent fraction we are looking for:

$$\frac{9.5}{100}\left[\frac{\times 90}{\times 90}\right] = \frac{\$855}{\$9{,}000}$$

The equivalent ratio, $\frac{\$855}{\$9{,}000}$, tells Jake that the \$855 tax Kyle paid was for a car that originally cost \$9,000.

Tips, Commissions, Fees, and Discounts as Percentages

Here are some other areas where you will encounter percent calculations in everyday life:

- **Tips and Gratuity:** Everyone likes receiving tips, but not as much as a hard-working server! Calculate the tip just as we did for sales tax, and remember, if you are looking for the final cost of your bill, you'll need to add that tip to your original bill.
- **Commission:** Commission is payment in the form of a percentage of how much an employee sells. Real estate agents, car salespeople, and some retail employees are just a few examples of jobs that get paid according to a commission percent. The more they sell, the more money they make! (Unlike gratuity, it's not added onto a customer's total bill; instead it is paid by the employer as payment for the employee's successful sale.) Commission is calculated the same way a tip is calculated.
- **Fee:** Percentage fees are charges for a service rendered based on a percentage of the service bill. Fees are different than commission in that they *are* paid by the customer. One type of fee might be what a personal injury lawyer would charge if she wins a lawsuit for a client. Typically that fee is around 30%–40% of the settlement (the money awarded to the plaintiff). So

if a $9,000 settlement is awarded to Julian in a court case, he would need to pay 30%–40% of that money to his lawyer.

- **Discounted Prices:** We love discounts as much as servers love tips! Stores often offer sales of 10%–75% off, so it's great to be able to see how much money you'll be saving. When answering a problem involving a percent discount, make sure you know if you are looking for the *amount of the discount* or the *sale price*. If you are looking for the sale price, you will need to calculate the amount of the discount in dollars, and then subtract that from the original price.

ERROR ALERT! It is a common error for students to think that a 20% discount means $20 off. If something cost $21 but was 20% off, would it be just $1 after the 20% discount? No way! Remember that 20% means 20 cents off of every 100 cents or $20 off every $100, so you must use your percent proportion to solve for the dollar amount of your discount.

Practice 1

Use the following scenario to answer questions 1 through 4:

Lulu is trying to decide between a pair of brown boots that cost $80 and a pair of black jeans that cost $70. She really wants them both, but she only has $85 in her wallet.

1. If tax is 5%, see if Lulu has enough money to buy the brown boots.
2. Lulu just found out that the store is having a 20% off sale on all footwear! How much would the pretax sale price be of the boots?
3. She was also just found out that all jeans are 30% off. Find the pre-tax sale price of the jeans.
4. Compare the prices and discounts on both items. Would the jeans or the boots be a better deal?
5. How is commission similar and different from gratuity?

Use the following scenario to answer questions 6 and 7:

Terry sells interior remodeling packages to clients looking to freshen up their homes. He gets a 25% commission on all of the packages he sells. He had a record-breaking month in March and made $11,500 in commission.

6. Will the $11,500 represent the part or the whole in the percentage proportion?

7. Calculate Terry's sales for the month of March.

8. Ming just won a lawsuit against a doctor who performed a major surgery on his wrong hand when Ming was under anesthesia. He won $36,000 but needs to pay his lawyer 40% of his settlement. How much money will be left over for Ming after he pays his attorney's fees?

Simple Interest

Depending on what you are doing with your money, you can either earn interest or pay interest. If you want to borrow money to buy a car, the bank will charge you a percentage of the money you borrow, which you will have to add to the money you pay back. If you invest your money into a special type of account, you will earn interest since you are allowing the bank to invest your money.

Let's look first at how banks calculate *Simple Interest* when you take out a loan. Simple interest is a percentage that is charged every year. So if you borrow money at a 10% interest rate for three years, you will need to pay 10% of the money you borrowed for each of those three years. Here is the simple interest formula:

Example: *Emile borrowed $6,500 to start a small business teaching kids competitive jump-roping and hopscotching. The interest rate was 8% and he will pay back the money in two years. How much money will he end up paying in interest?*

Solution:

Step 1: Use the percent ratio to calculate what 8% of $6,500 is by plugging in eight for the percent, 6,500 in for the whole, and finding the multiplication factor to make an equivalent fraction:

$$\frac{\%}{100} = \frac{part}{whole}$$

$$\frac{8}{100} = \frac{part}{6,500}$$

$$\frac{8}{100}\left[\frac{\times 65}{\times 65}\right] = \frac{520}{6,500}$$

So \$520 is 8% of \$6,500.

Step 2: Multiply your interest by the number of years.

Since Emile is borrowing this money for two years, multiply \$520 by two to determine the total interest he will pay: \$520 × 2 = \$1,040. Hopefully Emile has some great jump-roping skills because he is going to have to pay back an extra \$1,040 when he pays his \$6,500 loan back.

Whether you are calculating **simple interest earned or simple interest paid**, you must multiply the interest by the number of years that you are investing or borrowing money in order to calculate the total interest gain or loss.

Percent Increase and Decrease

Percent increase or decrease is one of the trickiest percent concepts for students. The most important thing to remember here is that you are comparing the *change in value*, to the *original value*. The change in value is always found by using subtraction. For example, if a shirt was originally \$20 and it is now \$15, you will be comparing the decrease in price, which is \$5, to the original price, which is \$20. The percent proportion for these questions will be:

$$\frac{\%}{100} = \frac{\textit{change in value}}{\textit{original value}}$$

Example: *Sienna is looking at a shirt that was originally \$20, and it is now \$15. Find the percentage that the shirt has been marked down.*

Solution:

Step 1: Write the Percent Increase/Decrease proportion:

$$\frac{\%}{100} = \frac{\textit{change in value}}{\textit{original value}}$$

Step 2: Calculate the change in value and plug that into the equation along with the original value. In this case the original value was \$20 and the change in value was \$5:

$$\frac{\%}{100} = \frac{5}{20}$$

Step 3: Find the multiplication factor to make an equivalent fraction (we will rewrite the $\frac{5}{20}$ on the left to show the multiplication more clearly):

$$\frac{5}{20}\left[\frac{\times 5}{\times 5}\right] = \frac{25}{100}$$

Since the ratio $\frac{5}{20}$ is equivalent to $\frac{25}{100}$, the percent decrease is 25%.

The same exact steps are used when the value of something has increased, like a stock or investment. We compare the value of the increase (found with subtraction) to the original value.

ERROR ALERT! With percent increase and decrease questions you are not comparing the *new value* to the *original value* in your percentage proportion! You are comparing the *change in value*, to the *original value*. The *change in value* is calculated with subtraction:

$$\frac{\%}{100} = \frac{\textit{change in value}}{\textit{original value}}$$

Practice 2

1. Owen gets $600 in cash for his 13th birthday from his grandparents and family members. His dad offers to pay him 7% simple interest if Owen lets his dad hold onto it until his 18th birthday. (Owen wants to buy himself a scooter on his 18th birthday.) If he decides to take his dad up on the offer, how much money will Owen have from this original $600 investment?

Use the following information to answer questions 2 through 4:

Owen wasn't impressed with how much money he was going to earn from his dad at 7% interest, so instead he invests his money into two different stocks. He invests $400 into KDL Solar Panels and $200 into a real estate investment firm, Sheryl Properties.

2. On his 18th birthday, the value of Owen's KDL Solar Panels stock is now worth $550. What was the percent increase on this investment over the five years. What was that percentage increase per year?

3. On his 18th birthday, the value of Owen's real estate stock is now $120. What was the percent decrease on this stock?

4. How much money in total did Owen gain through his two stock investments and how does that compare to what his dad had offered him?

5. If Micah puts $8,000 on a credit card that has a yearly interest rate of 15%, how much interest will she pay on that over the course of a year?

Answers

Practice 1

1. 5% tax on $80 boots will be $4, so Lulu has enough money to buy the brown boots.
2. 20% off of $80 will be $16 savings, so the sale price of the boots would be $64.
3. 30% off of the $85 jeans will be $25.50, so the jeans would cost $59.50.
4. Even though the jeans were originally $5 more expensive than the boots, with the higher percentage discount, the jeans are now about $5 cheaper, so they are the better deal.
5. Commission is pay that a salesperson receives from their employer or client for making a sale. Gratuity is a service charge that a customer pays for a service in addition to the original cost.
6. Since the $11,500 represents 25% of Terry's sales, it represents the *part* in the percentage proportion.
7. Terry sold $46,000 of interior remodeling packages to his clients in March.
8. Ming will have to pay his attorney $14,400 so Ming will have $21,600 left for himself.

Practice 2

1. 7% of $600 is $42 and since it would get this interest for five years the interest earned would be $42 × 5 = $210. Therefore this $600 would be $810 in five years.
2. Since his KDL Solar Panels stock investment went from $400 to $550 it's change in value was $150. An increase of $150 from $400 is 37.5%. 37.5% ÷ 5 years is equivalent to 7.5% per year.
3. Owen's real estate stock fell $80 in value, so compare that to its original value of $200 to determine the percent decrease. A drop of $80 from $200 is a 40% decrease.
4. Owen ended up making a total of $70 with his stock gains and losses. His dad's offer would have earned him $210, so that would have been a more lucrative investment.
5. 15% of 8,000 is $1,200 of additional interest Micah will have to pay.

14

Distributing and Factoring Linear Expressions

STANDARD PREVIEW

In this lesson we will cover **Standard 7.EE.A.1**. You will learn how to add, subtract, factor, and perform the distributive property on linear expressions with rational coefficients.

Operations on Linear Expressions

In Lesson 7 we saw that the properties of operations hold true for variables and learned how to add and subtract like terms. Remember that **like terms** are terms that have the exact same variable raised to the exact same power. We add or subtract *like terms* by performing the correct operation on the coefficients and keeping the variable exactly the same. For example, $10x + 8x = 18x$ and $9y - y = 8y$.

In this lesson we are going to expand our knowledge to operating on linear expressions with rational coefficients. A **linear expression** is a collection of constants and terms with variables that are joined together by addition or subtraction. Linear expressions don't have any exponents other than one. A rational number is any real number that can be written as a fraction or a terminating decimal. An example of a linear expression with rational coefficients is $\frac{1}{2}w + 10.8 + \frac{3}{2}w + 1.2$. We can simplify this problem by using the commutative property to rearrange the expression so that the like terms are together, and then we can combine them:

$$\frac{1}{2}w + 10.8 + \frac{1}{5}w + 1.2 =$$
$$\frac{1}{2}w + \frac{3}{2}w + 10.8 + 1.2 =$$
$$\frac{4}{2}w + 12 = 2w + 12$$

Adding Linear Expressions

When asked to add two linear expressions like $7q - 8$ and $12q + 5$, it is a good idea to rewrite any subtraction as addition by using *keep-switch-switch*. This will insure that you keep the negative signs with the correct terms:

Example: *Add the linear expressions* $7q - 8$ *and* $12q + 5$
Solution: Write this as a sum of two quantities: $(7q - 8) + (12q + 5)$
Since there are no factors outside the parentheses and addition is being performed, we can remove the parentheses: $7q - 8 + 12q + 5$
At this point there is sometimes confusion regarding which term the negative sign belongs to. To avoid confusion, turn the subtraction into addition by using *keep-switch-switch*: $7q + (-8) + 12q + 5$
Now you can safely add like terms $7q + 12q$ and $(-8) + 5$ to get $19q + (-3)$

Subtracting Linear Expressions

When asked to subtract two linear expressions, it is critical to set the expressions up with parentheses so that it's clear that *both* of the terms in the second linear expression will get subtracted. Since this concept can be very confusing for students, let's look at subtracting linear expressions in the context of removing fruit from a fruit basket:

A classroom fruit basket has nine bananas and eight apples. Oliver's teacher asks him, "Can you please remove five bananas and six apples from the basket for us to take to the park?" If Oliver follows her request, how many bananas and apples will now be in the fruit basket?

You can probably see that after Oliver follows his teacher's request by removing 5 of the 9 bananas and 6 of the 8 apples, there will be 4 bananas and 2 apples left in the fruit basket. Now let's frame this scenario using algebraic notation:

nine bananas and eight apples = $9b + 8a$
five bananas and six apples = $5b + 6a$
Now subtract what we are *removing* from what is *originally* in the basket:

$(9b + 8a) - (5b + 6a)$

Notice that we will have to subtract both of the terms in the second parentheses in order to arrive at our conclusion that there would be 4 bananas and 2 apples left in the basket. Subtracting both terms in the second parentheses is equivalent to giving the negative sign to both terms when rewriting the expression without parentheses:

$(9b + 8a) - (5b + 6a) =$
$9b + 8a - 5b - 6a =$
$\underline{9b} + \mathbf{8a} + \underline{-5b} + \mathbf{-6a} =$
$\underline{9b} + \underline{-5b} + \mathbf{8a} + \mathbf{-6a} =$
$4b + 2a$, which is 4 bananas and 2 apples

We hope this real-world example helped illustrate why the subtraction sign must get distributed to *both terms* in the second expression. Let's look at another subtracting example:

Example: *Subtract the linear expression* $4x + (-2)$ *from* $7x - 10$.
Solution: Write the quantity $7x - 10$ *first* since $4x + (-2)$ is being subtracted *from* it:

$(7x - 10) - (4x + (-2))$

When rewriting this without the parentheses, it is critical to distribute the negative sign to both the $4x$ and the -2 in the second parentheses:

$(7x - 10) - (4x + (-2)) = 7x - 10 - 4x - (-2)$

Since there is a lot of confusing subtraction, use *keep-switch-switch* to rewrite the subtractions as addition:

$7x - 10 - 4x - (-2) = 7x + (-10) + (-4x) + 2$

Now this equation is a lot more manageable. Combine like terms:

$\underline{7x} + (\mathbf{-10}) + \underline{(-4x)} + \mathbf{2} = 3x + (-8)$

ERROR ALERT! When subtracting linear expressions, a common mistake is for students to only subtract the first term in that expression. Make sure you initially write each linear expression in parentheses. In order to remove the parentheses you must subtract both terms in the second expression by changing *both* of their signs:

1. **Example:** Subtract $9c + 8$ from $3c + 5$
   Mistake: $3c + 5 - 9c + 8$ (Must write expressions in parentheses!)
   Correct: $(3c + 5) - (9c + 8)$
2. **Example:** Simplify $(6d - 1) - (4d + 5)$
   Mistake: $6d - 1 - 4d + 5$ (Must subtract the 5 as well!)
   Correct: $6d - 1 - 4d - 5$
3. **Example:** Simplify $(10p + 3) - (7p - 3)$
   Mistake: $10p + 3 - 7p - 3$ (Must subtract the 3 as well!)
   Correct: $10p + 3 - 7p - (-3) = 10p + 3 - 7p + 3$

You will now have a chance to practice adding and subtraction linear expressions with rational coefficients.

Practice 1

1. Add the linear expressions: $(\frac{1}{3}x + 5) + (\frac{1}{2}x + 9)$

2. What is $4.6x + 5$ subtracted from $9.4x - 3$?

3. Identify a common mistake that students might make with the following problem and explain why it would be wrong. Then perform the problem correctly: $(9x + 5) - (4x - 7)$

Use the following triangle to answer questions 4 through 6:

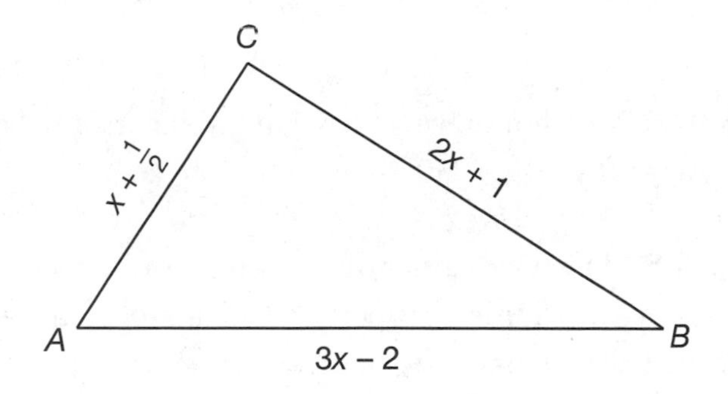

4. Find the perimeter of ΔABC.

5. Write, but do not solve, an expression that could be used to show how much longer $\overline{AB}$ is compared to $\overline{BC}$. (Hint: If the linear expressions are confusing you, make up side lengths for $\overline{AB}$ and $\overline{BC}$ and think about how you would use those side lengths to answer this question.)

6. Solve your expression from question 5.

Distributing with Linear Expressions

In Lesson 7 we learned that the Distributive Property is just as effective with variable expressions as it is with numbers. The **Distributive Property** for linear expressions states:

For any numbers a, b, and c: $a(b + c) = ab + ac$.

Before we move through an example, it is important to remind you that the Associative Property can be used to help us find the product of a constant and an algebraic term. Remember that the product $8(9v)$ can be rewritten as $(8 \times 9)v$ which simplifies to $72v$.

Example: *Use the Distributive Property to simplify* $6(7h + 5)$.

Solution: We can distribute the 6 by multiplying it to both terms inside the parentheses:

$$6(7h + 5) = 6(7h) + 6(5)$$

Then multiplying both terms by 6 will give us the final answer:

$$6(7h) + 6(5) = 42h + 30$$

ERROR ALERT! There are a handful of common errors to watch out for when using the distributive property with variable expressions. How do we know this? Because since the pencil was invented, these errors have regularly been scribbled onto paper! Rather than show you a handful of questions done correctly, we're going to show you what not to do when working with the distributive property! Carefully study the following pitfalls to avoid and the correct way to handle each challenge.

1. **When distributing a negative factor, the negative sign must go to both terms within the parentheses:**
   Mistake: $-8(r + 3) \neq -8r + 24$
   Correct: $-8(r + 3) = -8(r) + -8(3) = -8r + -24$
2. **Be careful when distributing a negative factor to a linear expression with subtraction:**
   Mistake: $-8(r - 3) \neq -8r - 24$
   Correct: $-8(r - 3) = -8(r) - (-8)(3) = -8r - (-24) = -8r + 24$
3. **Do not distribute multiplication over multiplication—the distributive property is only valid when a factor is being multiplied to a sum or difference!**
   Mistake: $8(3r) \neq 24 \times 8r$
   Correct: $8(3r) = 8 \times 3 \times r = 24r$
4. **PEMDAS is your copilot! Remember that the distributive property represents multiplication. Multiplication comes before addition and subtraction! You will find this popular mistake on Facebook with over ten thousand "Likes!"**
   Mistake: $2 + 3(6u - 7) \neq 5(6u - 7)$
   Correct: $2 + \underline{3(6u - 7)} = 2 + \underline{3(6u) - 3(7)} = 2 + 18u - 21 = 18u - 19$

5. **Watch out for subtraction in longer problems! Rewrite addition as subtraction to avoid incorrect distribution.**
Change $8(9f + 1) - \underline{3(6f - 7)}$ to $8(9f + 1) \underline{+ (-3)(6f + -7)}$ so that you are certain to distribute a (-3) to the second linear expression.
Correct: $8(9f + 1) \underline{+ (-3)(6f + -7)} = 8(9f) + 8(1) + (-3)(6f)$
$+ (-3)(-7)$ (We'll leave the rest to you!)

The distributive property works just as well with fractions and decimals as it does with whole numbers, so don't get too nervous as you work through the following practice problems.

Practice 2

1. $\frac{1}{4}(60y - 12) + 10$

2. What is the value of the expression $(\frac{2}{3} - 5x)$ after it's been tripled? *Hint: Think of how you would find the value of 5 after it's been tripled.*

3. Subtract the sum of $-7g$ and 8 from the linear expression $10.5g - 6\frac{1}{2}$. *Be careful with your signs!*

4. What are two different mistakes that a student could make when simplifying $10 - 4(2x - 3)$? Explain the mistakes and then illustrate step by step how to simplify this expression correctly.

5. Simplify the expression $\frac{4(5y - 3) - 12y + 4}{8y + (-8)}$

6. $0.1(80 - 10x) + \frac{1}{2}(12 - 6x)$

7. The formula for the perimeter of a rectangle is Perimeter = $2L + 2W$, where L = length and W = width. Given a rectangle where $L = 5x - 1$ and $W = 2x + 3$, what is the simplest expression that will represent the perimeter of this rectangle?

Factoring Linear Expressions

Similar to how *multiplication* is the opposite operation of *division*, *factoring* is the opposite property of *distributing*. While distributing requires us to *multiply* a common factor to two terms of a linear expression, *factoring* requires us to divide out the *greatest common factor* of two terms in a linear expression. Sure, we all love multiplication a little more than we love division, but factoring is here to stay and will only become more important in your future math classes, so it's a good idea to make friends with it now!

The **greatest common factor**, or **GCF** of two terms is the largest number that divides evenly into both terms. For example, 40, and 32 have a greatest common factor of eight since eight is the largest number that evenly divides both 40 and 32. If we reverse the order in the Distributive Property rule that was presented earlier in this lesson, it illustrates factoring:

For any numbers a, b, and c: $\underline{a} \times b + \underline{a} \times c = \underline{a}(b + c)$

Example: *Use factoring to rewrite* $40m + 32$ *as the product of two factors.*
Solution: Starting with $40m + 32$, identify the GFC as eight.
Rewrite $40m + 32$ as the sum of two products:
$40m + 32 = 8(5m) + 8(4)$
Now reverse the distributive property by pulling the GCF outside of a *single* set of parentheses (notice that multiplication could be used to change this expression back to its original form): $8(5m) + 8(4) = 8(5m + 4)$. Now $40m + 32$ is written as the product of two factors, 8 and $(5m + 4)$.
You can check your work by distributing. That should return you to your original expression: $8(5m + 4) = 8(5m) + 8(4) = 40m + 32$.

Practice 3

1. Factor out the GCF to create an equivalent expression to $(14x + 21)$.

2. Write two different equivalent expressions for $(-15 - 25k)$. Write one with a negative greatest common factor and write the second expression with a positive greatest common factor. Which factored form do you prefer and why?

3. How many common factors can you identify for $24n$ and 12? List all the different common factors for these two terms.

4. Use your answer from question 3 to write as many factored equivalent expressions to $24n - 12$.

5. Are $18 - 10x$ and $(9 - 5x)2$ equivalent expressions? Justify your answer.

6. Are $18 - 10x$ and $(-2)(-5x - 9)$ equivalent expressions? Justify your answer.

Answers

Practice 1

1. $(\frac{1}{3}x + 5) + (\frac{1}{2}x + 9) = \frac{5}{6}x + 14$
2. $(9.4x - 3) - (4.6x + 5) = 4.8x - 8$
3. A common mistake students might make would be to forget to give the negative sign to *both* terms in the second parentheses and rewrite it incorrectly as $9x + 5 - 4x - 7$. The correct way to rewrite this would be $9x + 5 - 4x + -7 = 5x - 2$.

Use the following triangle to answer questions 4 through 6:

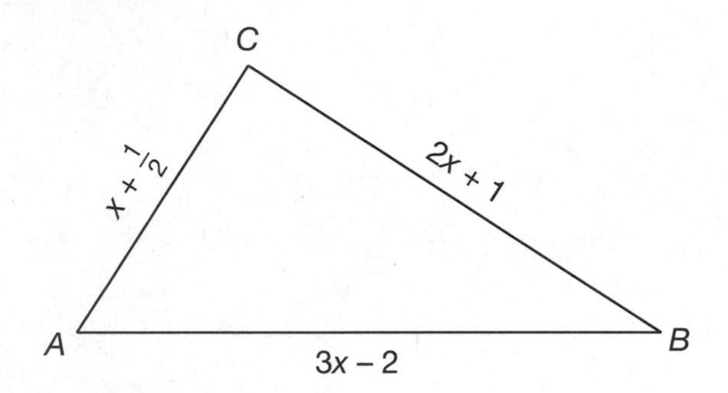

4. The perimeter of $\Delta ABC = 6x - \frac{1}{2}$
5. $(3x - 2) - (2x + 1)$ would show how much longer $\overline{AB}$ is compared to $\overline{BC}$.
6. $(3x - 2) - (2x + 1) = 1x - 3$

Practice 2

1. $\frac{1}{4}(60y - 12) + 10 = 15y - 3 + 10 = 15y + 7$
2. $3(\frac{2}{3} - 5x) = 2 - 15x$
3. $(10.5g - 6\frac{1}{2}) - (-7g + 8) = 17.5g - 14\frac{1}{2}$
4. $10 - 4(2x - 3)$ could be rewritten incorrectly the following ways:
   $10 - 8x - 3$ (forget to distribute the -4 to the 3)
   $10 - 8x - 12$ (forget to distribute the negative sign in -4 to the 3)
   $6(2x - 3)$ (combine the $10 - 4$ before doing the multiplication)
   Correctly done, $10 - 4(2x - 3) = 10 - 8x + 12 = 22 - 8x$ or $-8x + 22$
   ($8x - 22$ would be incorrect).
5. $\frac{4(5y - 3) - 12y + 4}{8y + (-8)} =$

   $\frac{20y - 12 - 12y + 4}{8y + -8} =$

   $8(\frac{1}{2})y - 20$

6. $0.1(80 - 10x) + \frac{1}{2}(12 - 6x) =$
$8 - 1x + 6 - 3x =$
$14 - 4x$

7. Perimeter = $2L + 2W$, $L = 5x - 1$ and $W = 2x + 3$
Perimeter = $2(5x - 1) + 2(2x + 3)$
Perimeter = $10x - 2 + 4x + 6$
Perimeter = $14x + 4$

Practice 3

1. $14x + 21 = 7(2x + 3)$

2. $-15 - 25k = -5(3 + 5k)$ and $-15 - 25k = 5(-3 - 5k)$. Although both answers are correct, it is slightly more conventional to write your quantity inside the parentheses as addition so it is useful to be able to write $-15 - 25k$ as $-5(3 + 5k)$.

3. Common factors of $24n$ and 12 include 1, 2, 3, 4, 6, and 12.

4. $24n - 12$ is equal to $2(12n - 6)$, $3(8n - 4)$, $4(6n - 3)$, $6(4n - 2)$, and $12(2n - 1)$

5. $18 - 10x$ and $(9 - 5x)2$ are equivalent expressions because the 2 on the right of $(9 - 5x)2$ is still indicating the distributive property, so $(9 - 5x)2 = 18 - 10x$.

6. $18 - 10x$ and $(-2)(-5x - 9)$ are not equivalent expressions because when the -2 is distributed to the $-5x$ it will become $10x$ and not $-10x$ as it is in the original expression.

15

Using Equivalent Expressions to Analyze Percents

STANDARD PREVIEW

In this lesson we will cover **Standard 7.EE.A.2**. You will first learn how to calculate percents by using multiplication. You will then learn to rewrite expressions in equivalent forms that help simplify percent increase and decrease problems. Lastly, you will learn how to rewrite expressions in different forms in order to deepen your understanding of the context of the situation.

Working with Percents Using Multiplication

In this lesson you are going to learn a new method for tackling percent questions: multiplication. In order to learn this method we must first review decimals and power of 10.

Writing Percentages as Decimals

In previous lessons we worked with percentages as part-to-whole ratios where the whole is 100. You now understand that 30% means 30 out of 100, which is written as $\frac{30}{100}$. Since all fractions can be written as decimals that means that percentages can also be written as decimals.

Multiplying and Dividing by Powers of 10

Powers of 10 are numbers that result from 10 being risen to a whole number exponent, such as 100 (10^2), 1,000 (10^3), and 10,000 (10^4). You probably are familiar with multiplying and dividing by powers of 10, but in case you need a quick refresher, let's review:

> A shortcut for *multiplying* a number by a power of 10 is to move the decimal point to the right, once for every zero in the power of 10: $6.5 \times 1{,}000 = 6{,}500$.
> Similarly, a shortcut for *dividing* a number by a power of 10 is to move the decimal point to the left, once for every zero in the power of 10: $6.5 \div 1{,}000 = 0.0065$. (Although .0065 is equivalent to 0.0065, it is always good form to write the optional zero before the decimal.)

Since percentages are fractions with denominators of 100, percentages can be easily converted into decimals by using the power of 10 shortcut. To represent any percentage as a decimal, simply move the decimal point of any percentage two spaces to the left and remove the percentage symbol.

$$30\% = \frac{30}{100} = 30 \div 100 = 0.30$$

$$5\% = \frac{5}{100} = 5 \div 100 = 0.05$$

$$120\% = \frac{120}{100} = 120 \div 100 = 1.20$$

$$8\tfrac{1}{2}\% = \frac{8.5}{100} = 8.5 \div 100 = 0.085$$

ERROR ALERT! When dealing with single-digit percentages, such as 5%, many students make a mistake with place value. Converting 5% to decimal form, it's common to write 0.5, rather than 0.05. Remember that 5% is 5 out of *100*, which is 5 *hundredths*, so it must be written as 0.05 and *not* 0.5.

Calculating Percentages with Multiplication

You may recall that the word *of* usually translates to multiplication. If Missy bought six boxes *of* 10 chocolates, we would know that she bought $6 \times 10 = 60$ chocolates. Since *of* means multiplication, and percents can be written as decimals, here is an alternate way to calculate percents:

Example: *If 30% of Missy's 60 chocolates are dark chocolates, how many dark chocolates does she have?*

Solution:

- First, translate 30% of 60 into a math expression: $\frac{30}{100} \times 60$
- Then rewrite 30% in decimal form: 0.30×60
- Last, perform the multiplication: $0.30 \times 60 = 18$, so Missy has 18 dark chocolates

This multiplication method with decimals can be used for finding sales tax as well as any other types of percentages.

Example: *Majadi is buying his dad a necktie that looks like a fish for Father's Day. The tie is $32 and sales tax is 6.5%. What will the total cost of the tie be?*

Solution:

- First, calculate how much the tax will be by calculating 6.5% of $32.
- Translate 6.5% of $32 into a math expression: $\frac{6.5}{100} \times 32$
- Then rewrite 6.5% in decimal form: 0.065×32
- Perform the multiplication to calculate the tax: $0.065 \times 32 =$ \$2.08
- Lastly, add the cost of the tax onto the price of the tie: \$2.08 + \$32 = \$34.08, so the total cost of the tie with tax will be \$34.08.

Practice 1

Use multiplication to solve the following percent problems.

1. What is 120% of a $600 investment?

2. What is 80% of 2 pounds?

3. What is $\frac{1}{2}$% of 3,000,000 people?

4. Dani is buying a travel package to Europe. The agency that is planning her tour charges a 4% booking fee for their services. If the trip she is purchasing is $1,200, what will the 4% fee be and how much will her entire booking cost?

5. 25% of the sophomore class at Osher University live off campus. If there are 3,200 students in the sophomore class, how many live off campus and how many live on campus?

6. Use rounding to approximate the sale price of a flat screen TV that is on clearance for 40% off if the original price is $595.

7. Frankie needs a new swimsuit for her upcoming competition. She is looking at a suit that is originally $60, but it's marked 20% off. Write out in words and numbers the steps for finding the sale price of this suit.

8. What was your last step for question 7? In the next section we're going to learn a way to approach percentage questions to eliminate this final step.

Writing Equivalent Expressions to Represent Percents

Did you notice that several of the questions in the previous practice set required two steps? When working with sales tax, price discounts, or service fees, until now you have had to calculate the amount of the percentage increase or decrease and then add or subtract that to the original amount to get your final answer. We know you have a lot of other interests outside of math, so we're now going to teach you a way to save time and approach these types of questions so there is just one calculation. In this section we will practice a new skill with variables and in the next section we will bring it to life with numbers.

First, let's begin by using variables to represent percentages:

What will a 25% discount be on a backpack that costs b dollars?
Multiply 25%, or 0.25, by b dollars: $0.25b$ will be the discount

What is a 150% increase of an initial investment of d dollars?
Multiply 150%, or 1.50, by d dollars: $1.5d$ will be the increase

What is $5\frac{1}{4}$% interest on a loan of p dollars?

Write $5\frac{1}{4}$% as 5.25% and multiply 0.0525 by p dollars: $0.0525p$

Next, let's consider how to write *a single expression* that could be used to calculate the sale price of that backpack that is 25% off in a single step. Let's begin by writing the equation out in words to show how the sale price is calculated:

Sale Price = Original Price – Discount

The original price is b. The 25% discount off b is $0.25b$. Substitute these values into the equation above:

Sale Price = $b - 0.25b$

Since b and $0.25b$ are like terms, we combine them. Remember that b is equivalent to $1b$ and when we combine like terms we just operate on the coefficients and keep the variables the same:

Sale Price = $1b - 0.25b = 0.75b$

Now we have two equivalent expressions that we can use to calculate the sale price of a backpack that is 25% off:

Sale Price = $b - 0.25b$ *or*
Sale Price = $0.75b$

The second expression representing the sale price, $0.75b$, shows that if we want to calculate the sale price of this backpack, we can do this in just one step by multiplying its original cost, b, by 0.75. We don't need to do the two steps of finding 25% of the backpack and then subtracting that from the original price. This is a concept that we can apply to other percentage problems:

When working with percentage discount, you can think of *p% savings* as being equivalent to paying (100 – *p*)% of the total cost of the item. For example, if something is 30% off, you will pay 70% of the item's original price.

Let's use the information from the previous box to consider this question: *What percentage of the total cost of an umbrella will you pay if it is on clearance for 75% off?* If an umbrella is marked 75% off, that means you will pay $(100 - p)$%, or 25%, of the original price. (This is just $\frac{1}{4}$ of the original price—so if an umbrella was normally $32, you would pay $8 if it were 75% off—that's a good deal!)

Example: *Write a single-term expression to represent the total amount Charles will pay for a dinner that has a bill of d dollars after leaving a 20% tip. Then use that single-term expression to calculate the total cost Charles will pay if he treats his friends to a $90 dinner and leaves a 20% tip.*

Solution:

- Write an equation out in words: *Total Cost = Original Bill + Tip*
- Original Bill = d and Tip = $0.20d$ so sub these values in:
  Total Cost = $d + 0.20d$
- Combine like terms to write as an equivalent expression:
  Total Cost = $1.20d$
- Since the expression $1.20d$ represents the total price with tip of a bill that originally cost d dollars, we can calculate the total cost by subbing $90 in for d:
  Total Cost = $1.20d = 1.20 \times \$90 = \$108$.

  By using the equivalent expression $1.20d$, we calculated Charles' total cost with just a single calculation. Thanks for dinner, Charles!

Practice 2

1. Is the expression $x + 0.08x$ representing a percent increase or decrease?

2. Use the previous expression, $x + 0.08x$, to determine the percent change that is being represented. Write a potential scenario for what this expression could be representing.

3. What type of percent change is being represented by the equation $p - 0.4p$?

4. Write an equivalent single-term expression to $p - 0.4p$ and explain a potential real-world situation this single-term expression could be used to determine an answer to.

5. An event planner is booking an art museum on a Saturday evening to host a fund-raising gala. The museum charges an additional weekend booking fee of 15% on top of the base price of services that cost d dollars. Write two equivalent expressions representing the total cost of the event including the additional weekend fee.

6. Use your single-term expression from question 5 to determine the total cost of a weekend event with a base price of $6,000.

7. Write a single-term expression to represent the sale price of a laptop that costs l dollars and is 10% off.

8. Use your answer from question 7 to calculate the sale price of a laptop that is $1,200.

Analyzing Equivalent Expressions

A useful aspect of equivalent expressions is they can be used to gain more information about a situation. In the following example we will see what kind of savings a customer has who buys multiple items at a percentage discount.

Example: Bella buys a book that is originally b dollars but is 20% off. Write two equivalent expressions representing the sale price of this book. Then write an expression representing how much Bella will spend if she buys *five* of these books for her closest friends. Evaluate what this final expression represents.

Solution:

- Write an equation out in words: Sale Price = Original Price – Discount
- Original price = b and discount = $0.20b$ so sub these values in:
  Sale Price = $b - 0.20b$
- Combine like terms to write as an equivalent expression:
  Sale Price = $0.80b$
- The two expressions representing the sale price of the book are $b - 0.20b$ and $0.80b$

Since Bella decides to buy 5 of these books, we can multiply both of the expressions $b - 0.20b$ and $0.80b$ by 5 to see how that changes our expressions:

$$5(b - 0.20b) = 5b - 1b = 4b$$

$$5 \times 0.80b = 4b$$

Since, after the 20% discount, Bella will spend $4b$ on 5 books that cost b dollars, we realize that with the sale, Bella is actually getting 5 books for the price of 4.

Sometimes discounts are given for items ordered in bulk. In this case, writing equivalent expressions can shine light on what the savings represent for larger orders.

Example: Sweet Jeanie's sells coconut macaroons for m dollars each. She prefers customers to buy at least a dozen macaroons, so she offers a 10% discount when people order 12 or more. Write an expression that represents the discounted price of a macaroon that was originally c dollars, but is now 10% off. Then determine an equivalent expression for a client who orders 50 macaroons for an office party. Analyze that expression.

Solution:

- Write an equation out in words: Bulk Discount Price = Original Price – Discount
- Original price = m and discount = $0.10m$ so sub these values in:

  Bulk Discount Price = $m - 0.10m$
- Combine like terms to write as an equivalent expression: Bulk Discount Price = $0.90m$
- In order to write an expression for a customer who is buying 50 macaroons, we can multiply both of the expressions $m - 0.10m$ and $0.90m$ by 50:

  $$50(m - 0.10m) = 50m - 5m = 45m$$

  $$50 \times 0.90m = 45m$$
- Since, after the 10% bulk discount, a customer will spend $45m$ on 50 macaroons, the customer is actually getting 50 macaroons for the price of 45. He is getting 5 free macaroons.

Practice 3

1. The demand for Tito's Tacos has increased so much that Tito has decided to increase his prices by 10%. Write a single-term expression that represents the new price of one of Tito's Tacos if the original price was t dollars.

2. Crosswoods School normally orders 200 tacos every Friday for their students and they are upset about the price increase. Use the single-term expression you wrote in question 1 to determine an expression that represents the current price of 200 tacos after the 10% increase.

3. Analyze the expression you wrote in question 2 to determine what the current cost of 200 tacos represents compared to the original price of tacos. How many tacos at the old price could have been purchased for the same price that 200 tacos now costs?

4. Beto's Burritos has been in competition with Tito's Tacos for years. Now that Tito's Tacos has increased their price, Beto's Burritos has decided to run a 5% discount on their burritos, which normally cost b dollars. Write a single-term expression that represents the new price of one of Beto's burritos after the 5% discount.

5. Crosswoods School decides to order 200 burritos from Beto's Burritos after they hear about Beto's 5% off promotion. Use the single-term expression you wrote in question 4 to determine a new expression that represents the current price of 200 burritos after the 5% discount.

Answers

Practice 1

1. 120% of \$600 = 1.2×600 = \$720
2. 80% of 2 pounds = $0.80 \times 2 = 1.6$ pounds
3. $\frac{1}{2}$% of 3,000,000 people = $0.005 \times 3{,}000{,}000 = 15{,}000$ people
4. 4% of \$1,200 = $0.04 \times$ \$1,200 = \$48, so her entire booking will cost \$1,248.
5. 25% of 3,200 = $0.25 \times 3{,}200 = 800$ students live off campus. 3,200 – 800 = 2,400 students live on campus.
6. 40% of \$595 ≈ 40% of \$600 = \$240, so \$240 is the discount. \$600 – \$240 = \$360, so the sale price will be \$360.
7. First find 20% of \$60 = $0.20 \times$ \$60 = \$12, this is the discount. Then subtract the \$12 discount from the original price: \$60 – \$12 = \$48.
8. The last step for question 7 was to subtract the sale price from the original price.

Practice 2

1. $x + 0.08x$ is representing a percent increase.
2. $x + 0.08x$ is representing a percent increase of 8%, so this could be the total cost of a toy dingo that costs x dollars and there is an 8% sales tax
3. $p - 0.4p$ is representing a decrease of 40%
4. A single term expression for $p - 0.4p$ is $0.6p$ and this could represent the price of a pineapple that is 40% off.
5. $x + 0.15x$ is one way this scenario can be represented. $1.15x$ is another expression that would just require a single step.
6. $1.15x = 1.15(\$6{,}000) = \$6{,}900$
7. $(100 - 10)\% = 90\%$, so $0.90l$ would be the expression used to determine the sale price of a computer that is 10% off.
8. $0.90l$, for $l = \$1{,}200$ is $0.90(1{,}200) = \$1{,}080$.

Practice 3

1. New taco cost = $1.1t$
2. New taco cost = $1.1t$, 200 tacos will cost $200 \times 1.1t = 220t$
3. Since 200 tacos now cost $220t$, Crosswoods School could have bought 220 tacos at the original price, t.
4. New burrito price = $0.95b$
5. New burrito price = $0.95b$, 200 burritos will cost $200 \times 0.95b = 190b$

16

Multi-Step Word Problems

STANDARD PREVIEW

In this lesson we will cover Standard 7.EE.B.3. You will learn how to approach and solve multi-step, real-life word problems.

Problem Solving 101

Students love to hate word problems. But if you think about it, word problems are really the most useful application of math, so it's a great idea to get over word-problem-phobia. After all, there won't be many days when your bill will come to your table at a restaurant and it will say "Solve for x if $\$140(1.2) = \frac{x}{5}$." What *will* happen though, is that you'll get a bill for \$140 and you'll wonder how to calculate the 20% tip and then how much each person from your group of five should chip in. Instead of hating word problems, look at them as an opportunity to put your hard-earned math skills to good use!

Students complain that word problems contain too many, um, *words*, and that they get confused over where to start. Here is a basic approach of how to tackle any word problem:

1. Read the question and underline or highlight any math words, numbers, or symbols in the problem.
2. Make sure you know what you are being asked to solve for. That question is often at the end of the problem. Circle it!
3. Now that you have your important information underlined and your question circled, organize your information in a chart, equation, or drawing and make a plan for how you are going to solve the problem.
4. Work step-by-step showing all your calculations along the way so that you can look back at them later to make sure there wasn't an error in your thinking or calculations.

Let's use these steps to walk through a few problems and techniques together.

Using Tables and Proportional Relationships

Example: Customers pay a $5 surcharge for a cab ride out of Denver International airport. On top of the airport fee, every mile costs $2.25. Laurel has $35 in her wallet and wants to be able to give her driver at least $3 as a tip. What is the farthest she can travel?

Solution:

1 & 2. Let's mark up the question and circle the question so that we can clearly see all the important parts: Customers pay a <u>$5 surcharge</u> for a cab ride out of Denver International airport. On top of the airport fee, <u>every mile costs $2.25</u>. Laurel <u>has $35 in her wallet and wants to be able</u> to give her driver <u>at least $3 as a tip</u>. (What is the farthest she can travel?)

3. Make a plan: Since Laurel has $35 and will have to pay the $5 airport surcharge and wants to leave at least a $3 tip, that means that $8 out of her $35 cannot get spent on mileage charges. Subtracting $8 from $35, we know that she has $27 to spend on mileage. We will see how many miles Laurel can travel by using proportional relationships in a chart.

4. Since we know that Laurel has $27 to spend on mileage, create a chart that shows the relationship between miles traveled and cost of mileage. We will start with 10 miles since we know that will cost $22.50. Then we will add $2.25 onto each additional mile until we have reached the farthest she can travel without surpassing $27:

Mileage Traveled	Mileage Cost at $2.25 per Mile
10 miles	10 × 2.25 = $22.50
11 miles	11 × 2.25 = $24.75
12 miles	12 × 2.25 = $27.00

We see that at 12 miles, Laurel will have $27 as a mileage cost, $3 for tip, and $5 for the airport surcharge, so that will use up all of her $35. She can travel up to 12 miles with the money she has.

Using Guess-and-Check Tables

When a proportional relationship doesn't fit the question, a table can be used for keeping track of your work and is a good place to organize your guess-and-check work as you search for the solution.

Example: Farhiyo and Asim sold T-shirts for a campus club last Saturday. They donated some of the money to a homeless shelter and put the remainder of the money into a scholarship fund that will be awarded to a graduating member of the senior class. They spent $210 printing the T-shirts and brought in $980 from the sale. If the club put $100 more into the scholarship fund than they donated to the homeless shelter, how much money did the club give to each cause?

Solution:

1 & 2. Let's mark up the question and circle the question so that we can clearly see all the important parts: Farhiyo and Asim sold T-shirts for a campus club last Saturday. They donated some of the money to a homeless shelter and put the remainder of the money into a scholarship fund that will be awarded to a graduating member of the senior class. They spent $210 printing the T-shirts and brought in $980 from the sale. If the club put $100 more into the scholarship fund than they donated to the homeless shelter, how much money did the club give to each cause?

3. Make a plan: Since the club spent $210 on supplies and brought in $980, that means that their profit was $770 after expenses ($980 – $210 = $770). This means that their donation to the shelter and their deposit into the scholarship fund must add up to $770.

4. Make a chart of the shelter donation, the scholarship fund money, and the total money earned. We will start with a guess for the shelter donation and then the scholarship fund money will be $100 more than that. The sum of these two numbers will represent how much the club would have had to earn in order to make those donations. Once we get the last column to total $770 that means we have our correct answers. We will use a guess-and-check method, which means for each row of the chart we reflect on our answer and then alter our next guess to get closer to the correct answer. Since $770 is close to $800, we will split $800 in half and use $400 as our first guess for the shelter donation and then fill out the other two columns accordingly:

Shelter Donation	Scholarship Fund ($100 More)	Donation + Scholarship ($770)
$400	$500	$900 (too high)

Notice that $400 to the shelter would mean that $500 went into the scholarship fund (since that donation was $100 more than the shelter donation), which would have required $900 profit from the T-shirt sales. Let's lower our starting guess to $300:

Shelter Donation	Scholarship Fund ($100 More)	Donation + Scholarship ($770)
$400	$500	$900 (too high)
$300	$400	$700 (too low)

Now our total donation amount of $700 is too low, so let's try $350 since that is halfway between our first two guesses:

Shelter Donation	Scholarship Fund ($100 More)	Donation + Scholarship ($770)
$400	$500	$900
$300	$400	$700
$350	$450	$800 (getting close)

You can see that our total is getting closer to the desired $770! A few more guesses will help us arrive at the correct answer:

Shelter Donation	Scholarship Fund ($100 More)	Donation + Scholarship ($770)
$400	$500	$900
$300	$400	$700
$350	$450	$800
$340	$440	$780 (just a touch too high)
$335	$435	$770 – Correct!

So, we were able to use a guess-and-check table to determine the club donated $335 to the shelter and $435 to their scholarship fund.

Practice 1

1. Meaghan's dance squad organizes a car wash in the municipal parking lot. It costs them $250 to rent the lot and they pay $35 for cleansers. If the squad charges $5 per car wash, how many cars must it wash to raise more money than its expenses?

2. Zelda is planting a garden this spring so she can have lots of fresh fruits and vegetables over the summer months. She plants five tomato plants that should yield 30 to 40 fruits each. She also plants two zucchini plants that yield 15 to 20 zucchinis each. Lastly, she plants four pepper plants that will grow five to eight peppers apiece. According to the yields of all of her plants, what are the least and most numbers of fruits and vegetables that Zelda will harvest this summer? *Hint: Make sure your table has one column for the plant name and quantity, one column for the low yields of each plant, and one column for the high yields of each plant.*

3. Austin and three of his college friends are driving from Ashland, OR, to Ojai, CA, for spring break. The round-trip distance will be 1,900 miles. Austin's car holds 12 gallons of gas and gets 384 miles per full tank of gas. If gas costs $4.40/gallon, and Austin and his friends will split the cost of gas evenly, what will each of them pay for gas?

4. Thea was surprised how easy it was to spend $25 going to the movies last weekend. Her ticket was $13.75 and she bought a large popcorn and soda. If the popcorn was twice as much as the soda, use a table to determine how much Thea spent on each of the popcorn and soda.

Drawing Pictures to Problem Solve

Have you ever heard the expression *a picture is worth a thousand words*? Sometimes, it really does pay to make an illustration to help work out a problem. Don't worry, we won't ask you to draw eyes full of shock or interlaced fingers, just some basic shapes, lines, and stick figures are all that's required for most math problems!

Example: There is a bare wall in the foyer of Mimi's home that is 12 feet long. She just purchased two pieces of artwork that are each $3\frac{1}{2}$ feet wide and she has asked you to hang them symmetrically, so that the distance between the two pieces of artwork is equal to the distance from the outside edge of each piece of artwork to the end of the wall on each side. Each painting will be hung from a single (durable) bracket that is directly in the middle of the back of the frame. How far from the edge of each wall will you hammer the nail into her wall, from which you will hang each painting? These are professionally painted walls with custom designer paint, so there is no room for error!

Solution: Pencil and paper please! Let's do a quick drawing that includes the dimensions of Mimi's wall and new pieces of artwork:

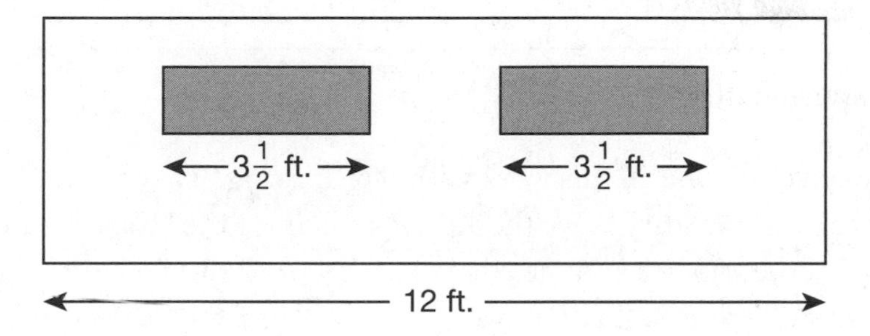

Ahh. . . . Once we have this drawn we start to feel a little better about the task at hand! We can see that the pictures are going to take up seven out of the 12 feet of wall space. This means that we have 12 – 7 = 5 feet of wall space for the gaps. Let's convert five feet to inches by multiplying five by 12: 5 × 12 = 60 inches. So there are 60" of free wall space that will be divided into three parts. $\frac{60''}{3} = 20''$ per gap. Let's add that to our drawing:

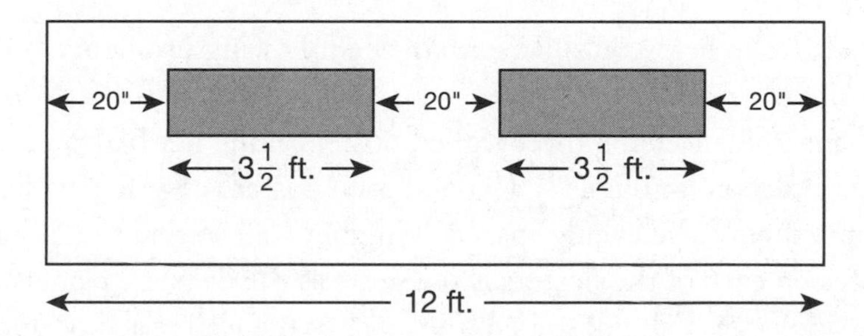

Next, we need to calculate where the bracket is going to line up along the wall. In order to do that we have to see how many inches the bracket is from the edge of the frame. Let's turn the $3\frac{1}{2}$ foot length of the artwork into inches and then divide that by two in order to find out where the middle of the frame is: 3.5 × 12 = 42 inches. $\frac{42''}{2} = 21''$, so the center bracket in the frame is 21" from its edge. Let's convert all our picture dimensions into inches and mark where the center bracket will be:

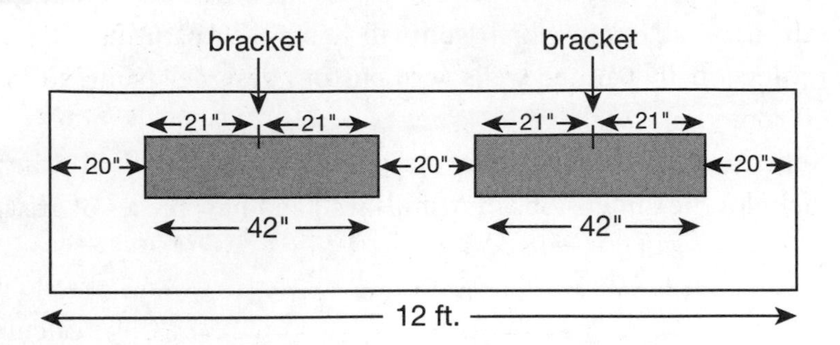

Now that we've done all this work with our drawing, we see that this problem is really not as scary as we thought it was! Since the bracket is 21" from the edge of the artwork and the edge of the artwork should be 20" from the end of the wall, we see that the two different nails for the brackets must be 41" from each end of the wall.

It's always important to check our work before doing anything permanent, so let's add all the inch lengths in our drawing to make sure they add up to 12 feet. Moving from left to right we have: 20" + 21" + 21" + 20" + 21" + 21" + 20" = 144". Since 12 feet times 12 inches gives exactly 144", we know that our drawing and calculations are correct. Hammer away with confidence!

Practice 2

Draw pictures to help you solve each of the following problems.

1. Pika is going to hang three signed posters along her bedroom wall. Her bedroom wall is $11\frac{1}{2}$ feet and each poster is $1\frac{1}{2}$ feet wide. She wants them to be evenly spaced along her wall so that the distance between each of the posters is the same as the distance from the outside edge of the posters to the edge of the wall. Each poster has a single bracket in the middle of it, from which it will hang. Determine where the three nails should go in her wall. Use a picture to support your work.

2. Belinda is making an urban garden on a plot of land donated by a business that does not use all of its green space. She is going to use the back fence along the property as one of the boundaries of her

garden and will use 50 feet of donated fence to create the boundary along the two widths and remaining length of fence. If she is going to make side lengths with the 50 feet of fencing using only whole-foot lengths, what dimensions will give her the most area for her garden? Use a drawing *and* a guess-and-check table to support your conclusion.

Checking Answers with Estimation

An estimate is just a rough approximation of a mathematical calculation. Being able to make fast and accurate mathematical estimates is one of the most important skills you can have in your personal life. Being able to estimate money can help you decide what you can and cannot afford when you're at the market. Sometimes, you might do fast calculations on your cell phone, but hitting a × instead of a + can cause you to pay a lot more money for something than you should. Estimating can also help you see if an answer you calculated makes real-world sense. The most straightforward way of estimating is to round your numbers before doing any operations.

Example: This week Ms. Katz worked for $3\frac{1}{2}$ hours on Monday, $6\frac{1}{3}$ hours on Tuesday, $5\frac{3}{4}$ hours on Wednesday, Thursday she was sick, and she worked $7\frac{1}{4}$ hours on Friday. If her pay is $12.35/hour, estimate her pay for this week.

Solution: Rather than adding all of the mixed numbers and then multiplying by the decimal, 12.35, we can round all of her work hours to see that she worked around 23 hours:

Day	Hours	Estimate
Mon	$3\frac{1}{2}$	4
Tues	$6\frac{1}{3}$	6
Wed	$5\frac{3}{4}$	6
Fri	$7\frac{1}{4}$	7
Total	?	23 hours

Now, instead of multiplying 23 hours by $12.35, round her hourly pay to $12 and perform 23 × 12 = $276. Ms. Katz should earn around $276 for her four days of work this week. (In case you are wondering, the *exact answer*, which took a lot more effort, is $281.58. As you can see, that is only around $5 off from the estimate, which we arrived at quickly and painlessly!)

Example: After eating a nice meal at Shutters, Kaoru is very happy with the service her family received and wants to tip her server at least 20%. If her bill was $78, do a fast calculation of how much tip Kaoru should leave.

Solution: There is a great trick for finding percentages that are multiples of 10%. First, round the number you are going to be taking the percentage of. In this case $78 rounds to $80. Then, find 10% of the number. Remember that 10% is $\frac{10}{100}$ or $\frac{1}{10}$ of the whole. Therefore, in order to find 10% of something we can divide it by 10. Dividing by 10 means moving the decimal point back to the left once. 10% of $80 = $8. 20% will be two times the value of 10%, so 20% is $16. Kaoru will leave a $16 tip.

Answers

Practice 1

1. Total expenses are \$250 + \$35 = \$285. At \$5/car, it will take them 57 cars to break even (break even means not lose money and not earn money, but be right at \$0). They will need to wash 58 or more cars if they want to make any money. They better have some good advertising!
2. Based on the following table, it seems like a low estimate is that Zelda will harvest 204 fruits and vegetables and a high estimate is 272 fruits and vegetables.

Name Number of Plants	Low Yield	High Yield
Tomato – 5 plants	5 × 30 = 150	5 × 40 = 200
Zucchini – 2 plants	2 × 15 = 30	2 × 20 = 40
Pepper – 4 plants	4 × 5 = 24	4 × 8 = 32

3. First find the unit rate for miles per gallon that Austin's car gets: 384 miles per 12 gallons is equivalent to $\frac{384}{12}$ = 32 miles per each gallon of gas. Set up a table representing gallons of gas and miles covered. Start with 1 gallon and 32 miles and then use multiplication relationships to get up to 1,900 miles.

Gallons	Miles
1	32
10 (10 × 1)	320 (10 × 32)
30 (3 × 10)	960 (3 × 320)
60 (2 × 30)	1,920 (2 × 960)

They will need about 60 gallons, which will cost a total of \$264, at \$4.40 per gallon. \$264 shared by four people will be \$66 per person.

4. Since her ticket was \$13.75 and Thea spent \$25 in total, she spent \$11.25 on popcorn and soda. Since the popcorn cost twice as much as the soda, set up a table with columns for soda, popcorn, and total and use guess-and-check reasoning until the total equals \$11.25.

Soda	Popcorn (Twice as much)	Total (Must be \$11.25)
\$5	\$10	\$15 (too high)
\$4	\$8	\$12 (very close)
\$3.75	\$7.50	\$11.25 (Correct!)

The soda cost \$3.75 and the popcorn cost \$7.50.

Practice 2

1. After turning all the measurements into inches and drawing a picture, we can see that Pika will have 84" of empty wall to be divided into 4 equal parts: $\frac{84}{4} = 21$". Since each poster is 18" wide, the hook is 9" from the end of it. Therefore, the nails for the end pictures will both be 30" from the edge of the wall and the center picture will have it's nail 69" from each wall (it will be in the middle of the wall).

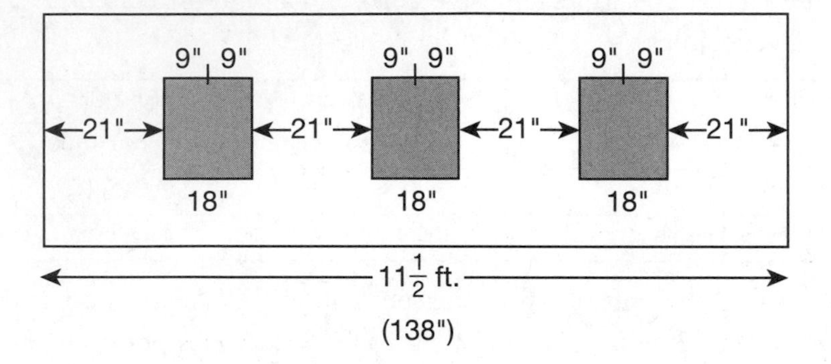

2. Belinda makes the following drawing to realize that her 50 feet of fence will have to get used as two widths and one length.

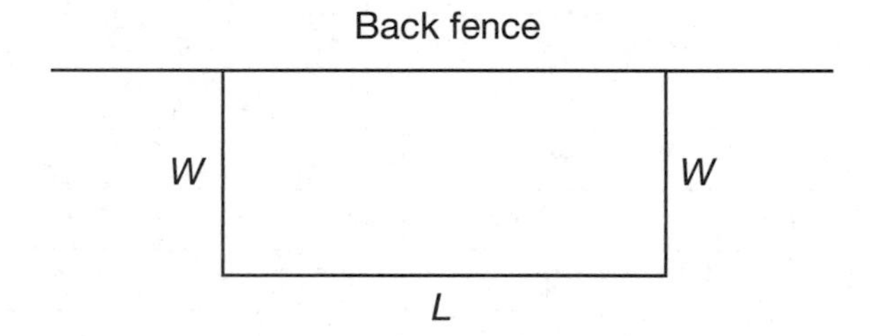

She sets up a table where she will input the width, double it, and then subtract that from 50 to see how much fence she'd have left over for her length. Then she finds the area by multiplying her length by width. Belinda starts with a width of 10 feet and increases her width by two inches until her area begins to decrease. She notices that the widths of 12 feet and 14 feet gave the two largest areas so her last effort is to try 13 feet for her width to see if that will result in a bigger area. She finds that the dimensions of 12×26 and 13×24 will both produce an equal area of 312 square feet for her garden, so she will use either of those dimensions.

Width	2 × Width	L (50 – 2W)	Area = $L \times W$
10	20	30	$10 \times 30 = 300$
12	24	26	$12 \times 26 = 312$
14	28	22	$14 \times 22 = 308$
16	32	18	$16 \times 18 = 288$
13	26	24	$13 \times 24 = 312$

17

Solving Two-Step Equations

STANDARD PREVIEW

In this lesson we will begin to cover Standards 7.EE.B.4.A and 7.EE.B.4.B. You will learn the difference between solving real-world problems with arithmetic versus solving them with algebra. You will also become proficient at solving equations that are in the form $px + q = r$ and $p(x + q) = r$, where p, q, and r are rational numbers. We will wrap up our work with these standards in the next lesson.

Two Approaches: Arithmetic versus Algebraic

In Lesson 5 we began learning how to recognize key words that can be translated into mathematical operations. We began our relationship with variables by expressing phrases like *five more than a number w* as an expression $5 + w$. In the next two sections we are going to build upon these skills by learning how to translate word problems into equations that can be solved algebraically. Let's look at an example of what we'll be covering:

Jackie works six-hour shifts at a boutique clothing store. If she earns $30 per shift and an additional 20% commission on all her sales, how much does Jackie need to sell in a single shift if she wants to make $114?

Although this question might look challenging, in the previous lesson we learned how to set up guess-and-check tables to answer questions like these. We solved questions using *arithmetic*, which involves only operations on numbers. Now we are going to learn to solve these types of questions by writing and solving equations using *algebra*, which means we'll use variables. Before we dive into Jackie's sales goal, let's compare the arithmetic approach to the algebraic approach for a simpler problem.

The perimeter of a rectangle is 48 inches. If its width is nine inches, what is the length of this rectangle?

We'll solve this at first using just arithmetic—no variables.

> **Step 1:** Since we know that the perimeter of a rectangle is two widths plus two lengths, we would double the width to see there are 18 total inches of width.
> **Step 2:** Then we'd subtract those 18 inches from 48 inches, to see that there are 30 inches of perimeter left for the remaining two lengths.
> **Step 3:** We would divide the remaining 30 inches by the 2 lengths to see that each length is 15 inches.

Next we'll use algebra to solve the exact same problem. Notice that we are performing the same steps in the identical order.

The perimeter of a rectangle is 48 inches. If its width is nine inches, what is the length of this rectangle?

Step 1: We identify that the perimeter formula is $P = 2L + 2W$	$P = 2L + 2W$
Step 2: We put our given information about the perimeter and width into the perimeter formula and simplify: $48 = 2L + 2(9)$	$48 = 2L + 2(9)$
Step 3: We simplify the 2(9)	$48 = 2L + 18$
Step 4: We subtract 18 from both sides.	$-18 \quad - 18$ $30 = 2L$
Step 5: We divide the remaining 30 inches by the 2	$\div 2 \quad \div 2$ $15 = L$ (Problem solved)

So as you see, the algebraic approach will give us the same answer, after following the same sequence of operations as the arithmetic approach. Sometimes it's fine to use only arithmetic to solve real-world problems, but in more complex situations the algebraic approach is more useful.

Reversing PEMDAS to Solve Two-Step Equations

If you were asked what value of x makes the equation $2x + 1 = 7$ true, you could probably arrive at the answer $x = 3$ without using algebraic steps to solve it. However, if you were given the problem $24x - 67 = 137$, it is unlikely that you could find a fast solution in your head. (We'll wait a moment as you try!) Let's establish a fixed set of steps that can be used to solve equations like this.

In Lesson 8 we tested algebraic equations by using PEMDAS to see if a given value made an equation true. We can see that $x = 3$ makes the equation $-8x + 6 = -18$ true by subbing three in for x and then following the order of operations as they appear in PEMDAS: $-8(3) + 6 = -18$. **Solving algebraic equations** is the opposite of this proccss: rather than being *given* a value for x to test an equation, we are *solving for* the specific value of x that makes an equation true.

Solving algebraic equations is like *undoing* them and therefore we must *reverse* the order of PEMDAS to arrive at our answer. We used opposite operations to solve one–step equations like $\frac{3}{2}x = 21$ or $x + 9.32 = 134.12$, but now the equations will get a little more involved. In 2–step equations, like $24x - 67 = 137$, we need to use addition or subtraction *and* multiplication or division to isolate x. The most important step to still remember when solving equations is that *what you do to one side of an equation, you must do to the other*!

Example: Reverse PEMDAS to solve for x in $24x - 67 = 137$

Solution:

Step 1: First rewrite subtraction as addition by using *keep-switch-switch* (this step is more useful in different types of equations, but it's good to get in this habit early).

Step 2: Reversing the order of PEMDAS, use the opposite operations of addition/subtraction to help get x alone. To cancel out a -67 we must add 67 to both sides.

Step 3: To undo the multiplication, divide both sides by 24 to get x alone

$24x + (-67) = 137$		(Step 1: keep-switch-switch)
$+67$	$+67$	(Step 2: addition/subtraction)
$24x$	$= 204$	(perform addition)
$\div 24$	$\div 24$	(Step 3: division/multiplication)
x	$= 8.5$	(Problem solved)

So the value $x = 8.5$ makes the equation $24x - 67 = 137$ a true statement. (This would have been tough to solve in our heads!)

ERROR ALERT! It is important to reverse the steps of PEMDAS and show your work clearly when solving an algebraic equation because there are certain mistakes that are too easy to make. For example, many students will see this equation and think they can immediately cancel out the eights: $3k + 8 = -8$. With careful inspection, you should see that after you subtract eight from both sides to get the $3k$ alone, you will have –16 and not zero on the right-hand side of the equation!

To solve all equations in the form $px +/- q = r$, for any rational numbers p, q, and r, follow these three steps to get x alone:

1. **Rewrite any subtraction to addition by using *keep-switch-switch***
2. **Use addition or subtraction (the opposite operation) to move q away from x**
3. **Use division to move p away from px and get x alone.**

Practice 1

1. Is the following statement true or false: Whether you are testing to see if a solution makes an equation true or solving an equation for a specific variable, it is necessary to follow the order of operations in PEMDAS. Explain your reasoning.

2. What is the difference between using arithmetic to solve a problem and using algebra to solve a problem? When is it more useful to use algebra?

3. What value of y makes the following equation true?
$18.30 + 3y = 52.80$

4. Solve for y in the equation $-14 = 5y + 5$

5. If $4p - 7\frac{3}{4} = 21.25$, what is the value of p?

6. Get the variable b alone: $rb - 13 = 13$ (Hint: your answer will have r in it.)

Why Use Keep-Switch-Switch?

Notice that we are actually listing three steps to solve "two-step equations." The first step, *rewriting subtraction as addition by using keep-switch-switch*, is particularly useful in equations where the variable term is being *subtracted from* a constant. Consider the equation $64 - \frac{1}{4}x = 144$. In equations like this, students are often tempted to add 64 to both sides of the equation because they see a minus sign:

$$\begin{array}{rl} 64 - \frac{1}{4}x = 144 & \\ \underline{+64 \qquad +64} & \text{WRONG!} \\ 128 - \frac{1}{4}x = 144 & \text{Noticc that the 64 didn't cancel out—it became 128!} \end{array}$$

This is incorrect since adding 64 did not help to get x alone! Using *keep-switch-switch* helps us see that the negative sign actually belongs to the $\frac{1}{4}x$ term. Once the negative sign is attached to the $\frac{1}{4}x$, it is easier to see that we must *subtract* the 64 to get rid of it on the left side of the equation:

$$\begin{array}{r} 64 + (-\frac{1}{4}x) = 144 \\ \underline{-64 \qquad\qquad -64} \\ -\frac{1}{4}x = 80 \end{array}$$

Now we have an equation where the variable is being multiplied by a fraction. No need to panic! Just recall that dividing by a fraction is the same as multiplying by it's reciprocal. Also, notice that the $\frac{1}{4}x$ is *negative*! Since we are trying to get x alone, and not $-x$ alone, we must be sure to divide by a $-\frac{1}{4}$ so that all that remains on the left side of the equation is x!

$$-\frac{1}{4}x = 80$$
$$\div -\frac{1}{4} \quad \div -\frac{1}{4}$$
$$x = \frac{80}{1} \times -\frac{4}{1}$$
$$x = -320$$

Notice that although the equation above had a fraction next to the variable, we solved it using the same steps. Once we got to the step $-\frac{1}{4}x = 80$, we divided both sides by $-\frac{1}{4}$ in order to get x alone, and since dividing by a fraction is the same as multiplying by it's reciprocal, $80 \div -\frac{1}{4}$ turned into $\frac{1}{80} \times -\frac{4}{1}$.

Solving Equations with Parentheses

Sometimes you will need to solve equations that are set up with parentheses. Let's consider the equation $4(c - 3) = 40$. Although you might be tempted to use the distributive property to get rid of the parentheses, there is a more efficient way to approach this problem. Notice that this equation is representing the *product* of 4 and the quantity $(c - 3)$. Do you remember what opposite operation we use to solve equations that involve products? We'll shade over the $c - 3$ to give you a hint:

$$4(c - 3) = 40$$
$$4(\blacksquare) = 40$$

Now, you might see that since *4 times* (■) equals 40, then the blacked out portion, (■), must equal 10, since $4 \times (10) = 40$. You are correct! (■) *does* equal 10. Since the expression $c - 3$ is really under the (■), we can conclude that $c - 3 = 10$. After adding three to both sides we see that $c = 13$. Of course all problems won't always have clear multiplication relationships that jump out at you like $4 \times 10 = 40$, so we will set out the steps for solving multiplication problems of a constant times a quantity here:

To solve equations in the form $p(x + q) = r$, where p, q, and r are rational numbers:

1. Divide both sides by p to get $(x + q)$ alone

2. Move q by using addition or subtraction (the opposite operation) to get x alone.

Example: Find the value of g that makes the equation true: $56 = -2.8(g + 10)$

Solution:

$$\begin{aligned} -2.8(g + 10) &= 56 \\ \div(-2.8) \quad & \quad \div(-2.8) \\ g + 10 &= -20 \\ -10 \quad & \quad -10 \\ g &= -30 \end{aligned}$$

Although you will still get the correct answer if you distribute the –2.8 first to the g and the 10, by using division as your first step, you are reducing the number of steps you have to take and reducing your room for error.

Checking Your Solutions

When solving equations, it's a good idea to check that your answer is correct. Do this by substituting the solution (the value you have found for the variable) into the original equation. If you follow the order of operations to simplify both sides of the equation, and you end up with a true statement, then you have the correct solution. For example, we can check if $x = 8.5$ is the correct solution to $24x - 67 = 137$:

$$\begin{aligned} 24x - 67 &= 137 \\ 24(8.5) - 67 &= 137 \\ 204 - 67 &= 137 \\ 137 &= 137 \checkmark \end{aligned}$$

Since both sides of the equation are the same, our solution of 8.5 is correct.

Practice 2

1. Use a first step of distributing $\frac{2}{3}$ and then solve for w in the equation: $\frac{2}{3}(w - 4) = 26$

2. Now divide by $\frac{2}{3}$ as your first step to solve for w in the equation: $\frac{2}{3}$ $(w - 4) = 26$

3. Compare your procedures and answers to questions 1 and 2. What method turned out to be easier?

4. Solve for e: $14.6 - 2.2e = -18.4$

5. Check your answer to question 4. If your solution is not correct, find your error.

6. Solve for x: $\frac{5}{6} = \frac{3}{4} - \frac{1}{12}x$

7. Check your answer to question 6. If your solution is not correct, find your error.

8. The following equation shows that Skye received \$5.75 in change after paying for three of the same T–shirts with a \$50 bill: $\$50 - 3t = \$5.75$. He can't find his receipt but he wants to know how much he was charged for each T–shirt.

Answers

Practice 1

1. False: When testing to see if a solution makes an equation true you must follow the order of operations in PEMDAS, but when solving an equation for a specific variable, it is necessary to *reverse* the order of operations in PEMDAS.
2. Arithmetic uses just the operations without writing equations with variables. The arithmetic approach might rely on guess-and-check tables or other visual methods to organize work. Algebra is useful in more complex situations with negative or fractional numbers or more complicated scenarios.
3. $y = 11.50$
4. $y = -3.8$
5. $p = 7.25$
6. $b = \frac{26}{r}$

Practice 2

1.
$$\frac{2}{3}(w-4) = 26$$
$$\frac{2}{3}w - \left(\frac{2}{3}\right)(4) = 26$$
$$\frac{2}{3}w - \frac{8}{3} = 26$$
$$+\frac{8}{3} \quad +\frac{8}{3}$$
$$\frac{2}{3}w = \frac{86}{3}$$
$$\div\frac{2}{3} \quad \div\frac{2}{3}$$
$$w = \frac{86}{3} \times \frac{3}{2}$$
$$w = 43$$

2.
$$\frac{2}{3}(w-4) = 26$$
$$\div\frac{2}{3} \quad \div\frac{2}{3}$$
$$(w-4) = \frac{26}{1} \times \frac{3}{2}$$
$$w - 4 = 39$$
$$+4 \quad +4$$
$$w = 43$$

3. The second method has less steps and fewer fractions in it, but both methods give the same answer.

4. $e = 15$

5. $14.6 - 2.2e = -18.4$, for $e = 15$

$$14.6 - 2.2(15) = -18.4$$
$$14.6 - 33 = -18.4$$
$$-18.4 = -18.4 \checkmark$$

6. $\frac{5}{6} = \frac{3}{4} - \frac{1}{12}x$

$$\frac{5}{6} = \frac{3}{4} + -\frac{1}{12}x$$
$$-\frac{3}{4} \quad -\frac{3}{4}$$
$$\frac{1}{12} = -\frac{1}{12}x$$
$$\div -\frac{1}{12} \quad \div -\frac{1}{12}$$
$$-1 = x$$

7. $\frac{5}{6} = \frac{3}{4} - \frac{1}{12}x$, for $x = -1$

$$\frac{5}{6} = \frac{3}{4} + -\frac{1}{12}(-1)$$
$$\frac{5}{6} = \frac{3}{4} + \frac{1}{12}$$
$$\frac{5}{6} = \frac{9}{12} + \frac{1}{12}$$
$$\frac{5}{6} = \frac{10}{12} \checkmark$$

8. $t = 14.75$, each T–shirt cost \$14.75

18

Real-World Equations and Inequalities

STANDARD PREVIEW

In this lesson we will finish covering **Standards 7.EE.B.4.A** and **7.EE.B.4.B**. You will learn how to model real-world problems with algebraic equations and inequalities, and you will use your skills from the previous lesson to solve them.

From English to Algebrish—Be Bilingual!

Algebrish really isn't a word, but we wanted to get your attention! You've practiced translating English into algebraic expressions and in this section you'll be writing and solving algebraic equations that represent real-world situations. Let's review some important words to watch out for:

is—*is* and most verbs will translate to an equals sign
and—*and* often means addition

less than/fewer than—remember that you'll need to switch the order of the items presented before you use subtraction with these two terms
of, each, per, every—these terms usually represent multiplication
twice, double—this term translates to ×2
triple—this term translates to ×3
unknown quantity—let whatever you're solving for be a variable

We'll start with a problem that models a similar scenario in Lessons 16 and 17. Remember that your first step should always be to write the mathematical relationship that exists between the working parts of your equation in words. Pay close attention to the four steps laid out.

Example: Jenna bought four identically priced nail polishes at the local pharmacy and paid with a $20 bill. She knows that she received $2.80 in change but cannot find her receipt. How much did each nail polish cost?
Solution:

Step 1: Read the question, underline the relevant parts, and circle what you are being asked to solve for.
Step 2: Determine how the parts of the problem relate to one another and write down the relationship between the parts in words and operations:

[Money paid] – [Cost of 4 nail polishes] = Change received

Step 3: Define your variables and given information:

n = cost of 1 nail polish
$4n$ = cost of 4 nail polishes
Money paid = $20
Change received = $2.80

Step 4: Replace the words in your equation from step 2 with the given numbers and your variable term:

$20 – $4n$ = $2.80

Now we have an algebraic equation that we can solve using the techniques learned in the previous lesson. Let's focus on translating a few more problems from words into algebraic equations before we begin solving them.

Percentage Word Problems

Example: Luke makes a consistent monthly salary. His employer puts $300 of his salary into a tax-free retirement savings account. After his retirement savings account, 20% of the remaining income gets deducted as income tax. If after these two deductions Luke's take-home pay is $2,800, what is his monthly salary before the retirement savings account deduction and income tax deduction?

Solution:

Step 1: Read the question, underline the relevant parts and circle what you are being asked to solve for.

Step 2: Determine how the parts of the problem relate to one another and write down the relationship between the working parts, using words and operations. Can you see where the 80% came from? (Since 20% of Luke's postretirement deduction salary is removed for tax, that means that 80% of it will be left over)

80% of (Monthly Salary – Retirement $) = [Take-Home Pay]

Step 3: Define your variables and given information:

***m* = Monthly Salary**
Retirement $ = $300
Take-Home Pay = $2,800

Step 4: Replace the words in your equation from step 2 with the given numbers and your variable term:

0.80(*m* – $300) = $2,800

Notice that in this problem it was really important to recognize that since Luke's employer takes 20% of his postretirement salary for tax, that is equivalent to paying him 80% of his postretirement salary.

Word Problems with Multiple Unknowns

Sometimes you will need to define more than one unknown in your equation in terms of the same variable. In the following equation we have three different unknowns. We must express them all using the same variable. You will see later that it is not possible to use three different variables for the three different people, because that would result in an equation that does not have a single unique answer.

Example: Abel, Bianca, and Caitlin are all going out to dinner. Abel makes the most amount of money so he offers to pay twice the amount that Bianca pays. Caitlin is a grad student, so the others agree that she can pay half as much as Bianca pays. If their total lunch bill is $42 including tax and tip, how much will they each pay?

Solution:

Step 1: Read the question, underline the relevant parts, and circle what you are being asked to solve for.

Step 2: Determine how the parts of the problem relate to one another and write down the relationship between all the working parts using words and operations.

[Abel] + [Bianca] + [Caitlin] = [Total Bill]

Step 3: Define your variables and given information. With this type of problem, since Abel is paying twice that of Bianca, and Caitlin is paying half that of Bianca, we let Bianca be b and write the other two people in terms of b. This step is critical!

Bianca = b
Abel = $2b$
Caitlin = $\frac{1}{2}b$

Step 4: Replace the words in your equation from step 2 with the given numbers and your three variable terms:

$$2b + b + \frac{1}{2}b = \$42$$

Step 5: Solve the equation by simplifying it to $3.5b = 42$ and then dividing both sides by 3.5 to arrive at $b = \$12$.

Step 6: You're not done yet! You only know what Bianca is going to pay! Answer the question by subbing $b = 12$ into each of the variable expressions that were written for each person:

Bianca = $b = \$12$
Abel = $2b = 2(\$12) = \$24$
Caitlin = $\frac{1}{2}b = \frac{1}{2}(\$12) = 6$

Step 7: We can check our answers by seeing that $\$12 + \$24 + \$6 = \$42$.

The only way to solve this type of question is to express all the different parts using *the same variable*. If we had defined the variables as Abel = a, Bianca = b, and Caitlin = c, we would have arrived at this equation, which would does not have one unique answer: $a + b + c = 42$.

ERROR ALERT! When needing to solve for more than one unknown in a word problem, it is a common error for students to assign a different variable to each unknown. It is instead necessary for each unknown to be represented with the same variable but in different algebraic terms, illustrating how the unknowns relate to each other. For example, $3x$ and x can be used if one unknown is three times larger than the other unknown. Another example could be the terms k and $7 + k$ to represent Kate's age and her friend's age, who is seven years older than Kate.

Practice 1

For each question, write your equation in words, define your variables, set up an equation, and solve it to answer the question.

1. Elliot is the personal assistant at a production agency. After finishing a big pitch, Elliot's team decides they want to celebrate with chocolate milkshakes. They send him to the Hubba Bubba Burger house where he buys six chocolate milkshakes. Elliot makes the milkshake maker laugh in the middle of her bad day, so she gives him a super-size upgrade on each shake for free. He tips her $1 for each shake. If the total he paid for the six shakes and tip is $31.50, how much did each shake cost before tip?

2. Sage has her mind set on a canyoncering harness that costs $80. She returned something last month and has a store credit of $139.09. She wants to use the entire credit now before she loses it, so she needs to pick something else out in the store. If a sales tax of 7% will be applied to her harness and other merchandise, and she wants her total purchase to be exactly $139.09, how much should she spend on her other merchandise?

3. Sammy, Lulu, and April Mae are competing for treats during an America's Top Terrier training session. April Mae couldn't follow instructions today, so she got five fewer treats than Lulu. Sammy was the cleverest of the bunch, so he got triple the amount of treats as Lulu. If the trainer gave out 35 treats, how many did each terrier get?

Inequalities

An inequality is like an equation but it has many different solutions, referred to as a *solution set*. We talked about an *inequality* earlier where a friend asked you to stop by his house "after 10:00 A.M." That would be represented as $h > 10$, which shows that 10.5, 11, or 12 are just some answers in the solution set. Conversely, $h = 10$ would be an *equation* representing your friend's request that you come over at exactly 10:00 A.M., and there is only one solution to that equation: 10:00 A.M. sharp!

Solving Inequality Equations

Although you're now a pro at solving equations, you haven't gotten your feet wet yet with solving inequalities. The general steps are identical, except there is one sneaky rule that promises to trick *every* student at least a few times! It's so devious that we're going to immediately highlight it here so that you pay special attention to it:

When solving inequalities in the form $px + q > r$ or $px + q < r$, where p, q, and r are rational numbers, it is necessary to switch the direction of the inequality symbol when dividing or multiplying by a negative number.

So this rule is stating that *when multiplying or dividing inequalities by a negative number*, a *less than* symbol will turn into a *greater than* symbol and vice versa. Whoa. That seems so strange! Let's take a look at $-6v > 30$ to see this confusing claim in action! First, we'll follow the rule and test a number from our solution set and then we'll ignore that odd rule and see what happens:

$$\begin{array}{r} -6v > 30 \\ \underline{\div(-6) \;\; \div(-6)} \\ v < -5 \end{array}$$

Since -10 is less than -5, let's sub -10 in for v and see if that makes the inequality true:

$$\begin{array}{r} -6(-10) > 30 \\ 60 > 30 \checkmark \end{array}$$

This is a true statement. Next, let's see what happens when we do not follow the rule:

$$\begin{array}{r} -6v > 30 \\ \underline{\div(-6) \;\; \div(-6)} \\ v > -5 \end{array}$$

Since zero is greater than negative five, let's sub zero in for v and see if that makes the inequality true:

$$-6(0) > 30$$
$$0 > 30 \times$$

Well, there we have it: if we do not switch the direction of the inequality sign when dividing by a negative, then we will get a solution set that does not work for the original equation. (Please note that this rule *only* applies to dividing or multiplying *by a negative number*. It doesn't get applied when dividing by a positive number and also does not get applied just because a final answer is negative.)

Inequality Word Problems

When you want to get *at least* an 80 on your history exam or your mom tells you to spend *a maximum of $15* on a birthday gift for a friend, you are entering the land of inequalities! There are countless real-world situations that don't have just one inflexible solution, and now we're going to express some of those situations as inequalities. Beware though, there is an additional step involved with inequalities that involves *intelligent rounding*. *What is intelligent rounding* you ask? *Intelligent rounding* is rounding that takes real-world context into consideration when rounding a number, rather than just following the rules you first learned to round numbers.

Intelligent Rounding

For example, if after some calculations you determine that you will need 1.4 gallons of paint for a room, can you just round that down to one and get the room painted with just one gallon? Of course not! The paint doesn't care that your elementary school teacher told you to round 1.4 down to one because it's closer to one than two. But that big bare spot on your wall is going to be really upset that you didn't do *intelligent rounding* when you were at the hardware store! In situations like this, you will need to intelligently round 1.4 gallons to two gallons since paint stores don't sell partial gallons of paint.

Example: Jim Smith works as boat captain on a whale watching boat called *The Harbor Club*. He gets paid $25 for a morning shift and an additional $3 for each paying client who books a tour. If Jim wants to make at least \$185 during his Monday morning shift, how many paying clients must book a tour on Monday?

Solution:

Step 1: Read the question, underline the relevant parts, and circle what you are being asked to solve for.

Step 2: Determine how the parts of the problem relate to one another and write down the relationship between the parts using words and operations. Since Jim wouldn't mind earning *more* than \$185, we face the *greater than* part of the inequality symbol toward the factors that will determine his pay and the *less than* part points toward the total pay of *at least \$185*.

[Shift Pay] + [Pay for Number of Clients] ≥ Total Pay

Step 3: Define your variables and given information:

c = Number of clients

\$3c = Subtotal Pay for c Clients

Shift Pay = \$25

Desired Total Pay = \$185

Step 4: Replace the words in your equation from step 2 with the given numbers and your variable term:

$\$25 + \$3c \geq \$185$

Step 5: Solve by subtracting 25 from each side and then dividing by 3:

$$\begin{array}{rr} \$25 + \$3c & \geq \$185 \\ -\$25 \qquad & -\$25 \\ \hline \$3c & \geq \$160 \\ \div \$3 & \div \$3 \\ \hline c & \geq 53\frac{1}{3} \end{array}$$

Step 6: The last step of solving an inequality is making sense of your answer. Since Jim needs at least $53\frac{1}{3}$ clients to make his goal, and he can't have $\frac{1}{3}$ of a client, we need to round this answer. We must round it up to 54 since 53 guests would have Jim making less than his desired \$185: $c \geq 54$.

Practice 2

1. What is a pitfall that many a student will make when solving the inequality $-3.5x < 70$?

2. Find the solution set to the inequality $-12 > 8 - 4x$

3. Use two numbers from your solution set in question 2 to check to see if they result in a true inequality. If they do not work, find your error.

4. Sydney is going to buy a bike at an annual clearance sale. The bikes are all 60% off and she has a $100 gift certificate that she can apply to the sale price of a bike. If Sydney lives in Bend, OR, where there is no sales tax and she doesn't want to spend over $750 after she applies her gift certificate, what would be the highest original price for a bike that she could consider?

5. Use a number from your solution set in question 4 to check to see if they result in a true inequality. If they do not work, find your error.

6. Francesca is going to Nicaragua. She sold some valuable stamps for $1,200 to pay for her round-trip airfare and lodging at a yoga retreat center. The yoga retreat center costs $60 per night and her round-trip plane ticket costs $550. What is the maximum number of nights she can stay at the retreat center if she doesn't want to spend more than her $1,200 stamp-sale earnings on her airfare and lodging?

7. How did you round your answer to question 6? Did you think about the context of the question? Find her total cost for lodging and airfare at the maximum number of nights you determined and make sure it is $1,200 or less.

Answers

Practice 1

1. $6 \times$ [Milkshake price] + [tip] = Total

   Milkshake price = m

   Tip = $3

   Total = $31.50

   $6m + 3 = 31.50$

   $6m = 28.50$

   $m = 4.75$, each milkshake cost $4.75.

2. Remember that a single-term expression can be used to find the total price of an item including a 7% tax: $0.07 \times$ (Subtotal) = Total with tax

   Now, the *subtotal* is going to be the harness plus an additional item:

   Harness = $80

   Additional item = i

   Subtotal = $80 + i

   And Sage wants the *total with tax* to be equal to her store credit of $139.09 so we write:

   $0.07 \times (80 + i) = 139.09$

   Instead of distributing 0.07, treat the quantity $(80 + i)$ as a single term and divide both sides by 0.07 to get:

   $80 + i = 129.99$

   $i = 49.99$, so Sage should buy something else for $49.99 if she wants to use up all of her available credit without spending any additional money.

3. We need to define all the dogs using the same variable. Since April and Sammy were both compared to Lulu, define all the dogs in terms of Lulu:

   Lulu = L
   April = $L - 5$
   Sammy = $3L$
   Lulu + April + Sammy = 35
   $L + (L - 5) + 3L = 35$
   $5L - 5 = 35$
   $5L = 40$
   $L = 8$
   Lulu = $L = 8$
   April = $L - 5 = 3$
   Sammy = $3L = 24$
   $8 + 3 + 24 = 35$, so we know our answer is correct!

Practice 2

1. They will forget to switch the direction of the inequality symbol when dividing by the –3.5.
2. $-12 > 8 - 4x$
   $-20 > -4x$
   $5 < x$, so x is greater than five
3. $-12 > 8 - 4x$, for $x = 6$
   $-12 > 8 - 4(6)$
   $-12 > 8 - 24$
   $-12 > -16$ ✔
4. Sydney wants the 60% off sale price to be less than the sum of $750 of her own money plus her $100 gift certificate:
   [60% off sale price] ≤ [$750] + [Gift certificate]
   Original Price = b
   The 60% off sale price will be equal to 40% of the original price of the bike:
   60% off sale price = $0.40b$
   gift certificate = $100
   $0.40b \leq \$750 + \$100$
   $0.40b \leq \$850$
   $b \leq \$2{,}125$, which means that if Sydney buys a bike that is less than or equal to $2,125 she will spend no more than $750 of her own money on her new bike. That's a great deal!

5. $b \leq \$2{,}125$, so we'll check \$2,100
$0.40b \leq \$750 + \$100$, for $b = \$2{,}100$
$0.40(\$2{,}100) \leq \$850$
$\$840 \leq \$850$ ✔

6. [Round-trip airfare cost] + [Lodging cost for n nights] ≤ [Stamp sale earnings]
Stamp sale earnings = \$1,200
Round-trip airfare = \$500
Yoga retreat center per night = \$60
Yoga retreat center for n nights = $\$60n$
$\$500 + \$60n < \$1{,}200$
$\$60n < \$700$
$n \leq 11.6$, so Francesca can only afford 11 full nights at the yoga retreat center with the budget that she has determined for herself.

7. Even though 11.6 rounded to 12 in elementary school, we must use intelligent rounding to realize that Francesca can only afford 11 nights at the yoga retreat center if she doesn't want to spend more on her airfare and lodging than the \$1,200 she earned. 12 nights at the retreat center would have cost 12(60) = \$720, which would sum to \$1,220 when added to her \$500 plane tickets.

19

Law of Exponents, Squares, and Cubes

STANDARD PREVIEW

In this lesson we will cover **Standards 8.EE.A.1** and **8.EE.A.2**. You will learn the properties of exponents and how to apply them to generating equivalent expressions. You will also learn about square and cube roots.

Law of Exponents

In Lesson 4 we learned that exponents serve as instructions for how many times a base will be multiplied by itself. We focused on numerical expressions, but exponents can be applied to variables as well. x^2, y^3, and πr^2 are a few examples of variables with exponents. Similarly to how the properties of operations helped us define rules for combining like terms ($4x + 5x = 9x$), there are six properties of exponents that are used to simplify exponential expressions. We will apply the properties of operations that you are already familiar with to uncover these laws. As we present each law, notice how each law can be uncovered by rewriting the exponential terms into expanded form and then applying simple operations. Each new law will be written in terms of variable bases and variable exponents in order to generalize it, but these laws work the same with numerical bases and exponents.

1. Multiplying Like Bases

What does $2^3 \times 2^4$ equal? It seems reasonable to think this will equal 2^{12} or 2^7. Write your guess here: ______. Now we'll figure out the rule by working through an example:

> In expanded form, $2^3 \times 2^4 = (2 \times 2 \times 2) \times (2 \times 2 \times 2 \times 2)$
> Since there are seven 2's being multiplied by each other in the expanded form, we can conclude that $2^3 \times 2^4 = 2^7$.
> Let's generalize this example: *when multiplying like bases, simply add the exponents and keep the base the same to create a simplified term.*

ERROR ALERT! Do not multiply the exponents when multiplying like bases! If you had 2^{12} as your guess, then you made this common mistake!

When multiplying like bases, $x^a \cdot x^b = x(a + b)$

Examples

1. $3^5 \cdot 3 = 3^5 \cdot 3^1 = 3^6$ (Notice that 3 is equivalent to 3^1)
2. If $x^6 \cdot x^4 = x^y$, then y must equal 10
3. If $2^2 \cdot 2^3 \cdot 2^4 = 2^x$, then $x = 9$

2. Dividing Like Bases

What does $2^6 \div 2^2$ equal? It seems reasonable to think it equals 2^3 or 1^3. Write your guess here: ______. Now we'll figure out the rule by working through an example:

> Let's write $2^6 \div 2^2$ in fractional form and then expand it:
> $$\frac{2^6}{2^2} = \frac{2 \times 2 \times 2 \times 2 \times 2 \times 2}{2 \times 2}$$

Since a number divided by itself always equals one, we can conclude that the first two pair of twos in the numerator will cancel out the pair of twos in the denominator:

$$\frac{\cancel{2} \times \cancel{2} \times 2 \times 2 \times 2 \times 2}{\cancel{2} \times \cancel{2}}.$$

Since we are now left with $2 \times 2 \times 2 \times 2$ we conclude that:

$$\frac{2^6}{2^2} = 2 \times 2 \times 2 \times 2 = 2^4.$$

Let's generalize this example: *when dividing like bases, subtract the exponents and keep the base the same to create a simplified term.*

ERROR ALERT! Do not divide the exponents or bases when dividing like bases! If you had 2^3 or 1^3 as your guess, then you made this common mistake!

When dividing like bases, $\frac{x^a}{x^b} = x^{(a-b)}$

Examples

1. $\frac{x^{10}}{x^6} = x^{(10-6)} = x^4$
2. $\frac{100^3}{100^2} = 100^{(3-2)} = 100^1 = 100$
3. If $\frac{7^9}{7^4} = 7^m$, then $m = 5$

3. Raising a Power to a Power

What does $(5^2)^3$ equal? It seems reasonable to think it equals 5^8 or 5^5 or maybe even 5^6. Write your guess here: ______ Now we'll figure out the rule by working through an example:

Let's write $(5^2)^3$ in expanded form by writing 5^2 multiplied by itself three times:

$$(5^2)^3 = (5^2) \times (5^2) \times (5^2)$$

Now we'll expand it one more time to see that 5 is being multiplied by itself 6 times:

$$(5^2)^3 = (5 \times 5) \times (5 \times 5) \times (5 \times 5) = 5^6$$

Let's generalize this example: *when a base with a power is raised to another power, multiply the exponents and keep the base the same to create a simplified term.*

ERROR ALERT! Do not raise the inside exponent to the power of the outside exponent when raising a power to a power! If you had 5^8 as your guess, then you made this common mistake!

When raising a power to a power, $(x^a)^b = x^{(a \times b)}$

Examples

1. $(5^4)^5 = 5^{20}$
2. If $(0.1^7)^7 = 0.1^m$, then $m = 49$
3. If $(n^2)^4 = n^m$, then $m = 8$

4. Raising a Product to a Power

What does the product $(2x)^3$ equal? It seems reasonable to think it equals $2x^3$. Do you have any other ideas of what it could be? You may have gotten tricked by the last few rules, so think hard! Write your guess here: ______ Now let's figure out the rule by working through an example:

Let's write $(2y)^3$ in expanded form by writing $2y$ multiplied by itself three times:

$$(2y)^3 = (2y) \times (2y) \times (2y)$$

Now use the Associative and Commutative Properties to rearrange the terms so that there are three 2's and three y's being multiplied:

$$(2y)^3 = (2 \times 2 \times 2) \times (y \times y \times y) = 2^3y^3$$

Do you notice how the power of 3 didn't just go to the y? The power of 3 also went to the 2 since the 2 was inside the parentheses. Let's generalize this example: *when a product is raised to a power, raise each factor inside the parentheses to the power outside the parentheses.*

ERROR ALERT! Do not just give the exponent outside the parentheses to the variable part(s) inside the parentheses. If your guess was $2x^3$, then you made this common mistake!

5. Raising Terms to a Power of Zero

What does $(8)^0$ equal? It seems reasonable to think it equals zero or maybe eight. Do you have any other thoughts? Write your guess here: ______ Let's figure this tricky rule out by returning to our law for dividing like bases:

Since exponents are subtracted when like bases are divided let's start with the following equation where we have randomly chosen the exponent two for the numerator and denominator (you will see that we could have chosen any exponent to arrive at this conclusion.)

$\frac{8^2}{8^2} = 8^{(2-2)} = 8^0$

The quotient on the left equals 1: $\frac{8^2}{8^2} = \frac{8 \times 8}{8 \times 8} = \frac{64}{64} = 1$. (When the exponent is the same in both the numerator and denominator, this will always be the case since anything divided by itself is always one.)

$(\frac{8^{12}}{8^{12}} = 1$ or $\frac{8^5}{8^5} = 1$, etc.)

Therefore, combining our first conclusion that $\frac{8^2}{8^2} = 1$ and our second conclusion that $\frac{8^2}{8^2} = 8^0$ we can use substitution to show that 8^0 is equivalent to one.

$8^0 = 1$

We imagine this is kind of weird for you, but raising something to the power of zero *does not mean* multiplying it by zero! This is one of the trickiest rules for students to remember!

Let's generalize this example: *any base raised to a power of zero equals one.*

ERROR ALERT! Do not confuse that something raised to the power of zero equals zero! This is the most common error students make using this property!

....

When raising a term to zero, $(x)^0 = 1$

....

Examples

1. $99^0 = 1$
2. $(3m)^0 = 1$ (The $3m$ is a single term, so there's no need to give the 0 power to both the 3 and the m.)
3. $5cd^0 = 5c \times d^0 = 5c$. (Notice that the zero *only* belongs to the d in this term.)

6. Negative Exponents

What does $(5)^{-2}$ equal? It's tempting to think that $(5)^{-2}$ is equivalent to -10, right? Write your guess here: ______ Let's figure this sneakiest rule out by returning to our law for dividing like bases:

Since exponents are subtracted when like bases are divided let's start by considering the following problem, where we've randomly chosen 3 and 5 to be our exponents since they subtract to get a –2 exponent:

$$\frac{5^3}{5^5} = 5^{(3-5)} = 5^{-2}$$

Now let's write $\frac{5^3}{5^5}$ in expanded form:

$$\frac{5^3}{5^5} = \frac{5 \times 5 \times 5}{5 \times 5 \times 5 \times 5 \times 5}$$

Three sets of fives in the numerator can now be canceled out with three sets of fives in the denominator. This will leave a one in the numerator and two fives in the denominator:

$$\frac{5^3}{5^5} = \frac{\cancel{5} \times \cancel{5} \times \cancel{5}}{\cancel{5} \times \cancel{5} \times \cancel{5} \times 5 \times 5} = \frac{1}{5 \times 5} = \frac{1}{5^2}$$

Since our first expression shows that $\frac{5^3}{5^5} = 5^{-2}$ and our expanded form shows that $\frac{5^3}{5^5} = \frac{1}{5^2}$, using substitution shows that it must be true that 5^{-2} is equivalent to $\frac{1}{5^2}$. We hope that doesn't make you feel *too* uncomfortable. We don't want you to just take our word for it, so reread the previous work until you can accept that bases with negative exponents don't mean negative answers, but instead will result in the reciprocal of a base.

Let's generalize this example: *a base raised to a negative power is equivalent to the reciprocal of that base with the positive value of the power.* $3^{-2} = \frac{1}{3^2}$

ERROR ALERT! Do not get tricked into thinking that a base raised to a negative power equals a negative number! Students all over the world make that error with this property!

When dealing with negative exponents, $x^{-a} = \frac{1}{x^a}$

Examples

1. $2^{-3} = \frac{1}{2^3} = \frac{1}{8}$
2. $5^{-2} = \frac{1}{5^2} = \frac{1}{25} = \frac{4}{100} = 0.04$ (Notice that 5^{-2} becomes 0.04. That was unexpected!)
3. $3x^{-2} = \frac{3}{x^2}$ (Notice that the negative exponent only belongs to the x so only the x goes in the denominator and the three stays in the numerator.)

4. $(3x)^{-2} = \frac{1}{(3x)^2} = \frac{1}{3^2x^2} = \frac{1}{9x^2}$ (Notice that with the parentheses, the negative exponent belongs to the entire $3x$ term, so the entire term moves to the denominator. Then the *product to a power rule* is used to give the exponent of two to both the three and the x.)

Practice 1

1. What is the value of k if $5^4 \times 5^2 = 5^k$?

2. What does it mean when a number is raised to the third power? Write $(\frac{2}{3})^3$ in expanded form and write an equivalent expression for $(\frac{2}{3})^3$.

3. Write $(3b)^4$ in expanded form and simplify it into an equivalent expression.

4. What does a negative exponent mean? What fraction, decimal, and percentage does 2^{-2} represent?

5. What does it mean when a variable doesn't have an exponent, like v. Is the exponent of v just zero?

6. Use your answer to question 5 to simplify the expression $g^5 \times g$.

7. What does an exponent of zero mean? What is different between the values of the expressions $10k^0$ and $(10k)^0$?

8. Can you use any of the laws of exponents to simplify the product of $4^2 \times 3^3$? Would it be 12^5, or 12^6, something else, or not possible to rewrite as a single term?

9. Find the value of $(\frac{1}{4})^{-2}$ by following the rules for negative exponents and the rules for fractions.

Square Roots and Cube Roots

Similar to how addition is the opposite of subtraction and multiplication is the opposite of division, roots are the opposites of exponents. Roots are the tools used to *undo* exponents. We are going to focus on *square roots* and *cube roots* in this lesson, but as you move into higher levels of math you will learn about many different kinds of roots.

Square Roots

Square rooting involves backward thinking. For instance, if a problem reads, "What is the square root of 16?" it is asking you for the number that equals 16 when multiplied by **itself**. The square root symbol looks like an elongated check mark and the number that is being square rooted goes into the box. If asked to *take the square root of 16*, that task would be represented like this:

$$\sqrt{16}$$

In order to simplify a square root, we must find the number that, when multiplied by itself, is 16. So when looking at $\sqrt{16}$ the question we are trying to answer is *what number equals 16 when it is squared*? Since we are undoing the multiplication, we are using division. The answer is four!

$$4 \times 4 = 16$$
$$4^2 = 16$$
$$\text{So, } \sqrt{16} = 4$$

ERROR ALERT! A common error for students to make is to think that taking the square root of a number means dividing it by two. $\sqrt{16} \neq 8$ since $8 \times 8 = 64$ and not 16!

Cube Roots

Cube roots function similarly to square roots. To find a cube root of a number, ask yourself, "What number when multiplied by itself three times gives me this number?" Mathematically, cube roots are written like a square root, but with a 3 in the corner: $\sqrt[3]{-27}$. So, what is an equivalent expression to $\sqrt[3]{27}$? In order to solve this we need to find the number that when multiplied by itself three times yields 27. Three!

$$3 \times 3 \times 3 = 27$$
$$3^3 = 27$$
$$\text{So, } \sqrt[3]{27} = 3$$

ERROR ALERT! A common error for students to make is to think that taking the cube root of a number means dividing it by 3. $\sqrt[3]{27} \neq 9$ since $9 \times 9 \times 9 = 729$ and not 27!

Perfect Square, Perfect Cubes, and Irrational Numbers

When a number has a square root that is an integer, we say that number is a **perfect square**. Numbers like 4, 9, and 25 are all perfect squares because $\sqrt{4} = 2$, $\sqrt{9} = 3$, and $\sqrt{25} = 5$. When a number has a cube root that is an integer, we say that number is a **perfect cube**. Numbers like 8, 27, and 1,000 are all perfect cubes because $\sqrt[3]{8} = 2$, $\sqrt[3]{27} = 3$, and $\sqrt[3]{1,000} = 10$. A **rational number** is any number that can be expressed as a fraction or has a terminating or repeating decimal. Conversely, an **irrational number** is a number that cannot be written as a fraction and has non–repeating decimals that never end.

Entering the land of roots means entering the land of irrational numbers! When the square root is taken of any number that is not a perfect square, the result is an irrational number. Similarly, when we take the cube root of a number that isn't a perfect cube, it also produces an irrational number. Therefore, $\sqrt{2}$, $\sqrt{8}$, $\sqrt{10}$ and $\sqrt[3]{4}$, $\sqrt[3]{9}$, and $\sqrt[3]{15}$ are all irrational numbers since none of these are perfect squares or cubes. It is helpful to be familiar with the Perfect Squares and Perfect Cubes in the following table:

Base	Perfect Square	Perfect Cube
1	1; (1^2)	1; (1^3)
2	4; (2^2)	8; (2^3)
3	9; (3^2)	27; (3^3)
4	16; (4^2)	64; (4^3)
5	25; (5^2)	125; (5^3)
6	36; (6^2)	216; (6^3)
7	29, (7^2)	You won't really see this!
8	64; (8^2)	You won't really see this!
9	81; (9^2)	You won't really see this!
10	100; (10^2)	1,000; (10^3)

Square and Cube Roots of Fractions

Although it looks really intimidating, taking the square root or cube root of a fraction is the same thing as taking the root of the numerator and denominator. For example, $\sqrt{\frac{1}{4}}$ is the same expression as $\frac{\sqrt{1}}{\sqrt{4}}$. We can see that $\frac{\sqrt{1}}{\sqrt{4}}$ simplifies to $\frac{1}{2}$.

Similarly, $\sqrt[3]{\frac{8}{27}}$ is equivalent to $\frac{\sqrt[3]{8}}{\sqrt[3]{27}}$ which simplifies to $\frac{2}{3}$.

Square & Cube Roots of Negative Numbers

The final characteristics we are going to discuss about roots is taking the square or cube root of negative numbers. It is never possible to take the square root of a negative number since two identical numbers will never have a negative product. (A *positive* × *positive* = *positive* and a *negative* × *negative* = *positive*.) Therefore, $\sqrt{-9}$ has no solution and we say that it is *not a real number*.

Watch out! The same is not true for cube roots though! Since a negative multiplied by itself *three* times produces a negative number, it is possible to take the cube of a negative number. Notice that $-3 \times -3 \times -3 = -27$, so therefore, $\sqrt[3]{-27} = -3$.

Practice 2

1. What is the difference between $\sqrt{1{,}000}$ and $\sqrt[3]{1{,}000}$? What is the simplified value of each expression?
2. What is the value of $\sqrt[3]{125} + \sqrt{49}$?
3. What is the value of $\sqrt{\frac{25}{36}}$?
4. Simplify $\sqrt[3]{-64}$.
5. Can $\sqrt{0}$ and $\sqrt[3]{0}$ be simplified? What about $\sqrt{1}$ and $\sqrt[3]{1}$?
6. What is the difference of the largest double-digit perfect square minus the largest double-digit perfect cube?
7. Which of the following number(s) is/are a perfect square *and* a perfect cube: 1, 8, 36, 64, 100.

Answers

Practice 1

1. If $5^4 \times 5^2 = 5^k$, then $k = 4 + 2 = 6$.
2. Raising a number to the third power means multiplying it by itself three times, so $(\frac{2}{3})^3 = (\frac{2}{3}) \times (\frac{2}{3}) \times (\frac{2}{3}) = \frac{2^3}{3^3} = \frac{8}{27}$
3. $(3h)^4 = 3h \times 3h \times 3h \times 3h = 3^4h^4 = 81h^4$
4. A negative exponent means that the base is moved into the denominator and the power becomes positive. $2^{-2} = \frac{1}{2^2} = \frac{1}{4} = \frac{25}{100} = 0.25 = 25\%$. That's a lot of different forms for just one exponent!
5. When a variable doesn't have an exponent indicated, its exponent is 1: $v = v^1$.
6. $g^5 \times g = g^5 \times g^1 = g^6$.
7. Anything raised to an exponent of zero equals 1. $10k^0$ and $(10k)^0$ mean two different things: With $10k^0$, only the k has the exponent of 0 so, $10k^0 = 10 \times k^0 = 10 \times 1 = 10$. But with $(10k)^0$, the $10k$ is in a set of parentheses so the entire $10k$ term is raised to the power of zero and will equal 1: $(10k)^0 = 1$.
8. None of the laws from this lesson can be used to simplify $4^2 \times 3^3$ into a single term.
9. $(\frac{1}{4})^{-2} = \frac{1}{\left(\frac{1}{4}\right)^2} = \frac{1}{\left(\frac{1}{4}\right)\left(\frac{1}{4}\right)} = \frac{1}{\left(\frac{1}{16}\right)} = 1 \times (\frac{16}{1}) = 16$

Practice 2

1. $\sqrt{1{,}000}$ is asking "What number times itself two times equals 1,000?" $\sqrt[3]{1{,}000}$ is asking "What number times itself three times equals 1,000?" $\sqrt{1{,}000}$ is not a perfect square and therefore is an irrational number. $\sqrt[3]{1{,}000}$ is a perfect cube and equals 10.
2. $\sqrt[3]{125} + \sqrt{49} = 5 + 7 = 12$
3. $\sqrt{\frac{25}{36}} = \frac{\sqrt{25}}{\sqrt{36}} = \frac{5}{6}$
4. $\sqrt[3]{-64} = -4$ since $-4 \times -4 \times -4 = -64$
5. $\sqrt{0} = 0$ and $\sqrt[3]{0} = 0$. $\sqrt{1} = 1$ and $\sqrt[3]{1} = 1$
6. The largest double-digit perfect square = 81. The largest double-digit perfect cube = 64. The difference between them is 17.
7. One and 64 are the only numbers less than 100 that are both a perfect square *and* a perfect cube.

20

Scientific Notation

STANDARD PREVIEW

In this lesson we will cover **Standards 8.EE.A.3** and **8.EE.A.4**. You will learn how to write extremely large and extremely small numbers in scientific notation. You will also learn how to perform operations on numbers expressed in scientific notation.

Basics of Scientific Notation

Scientific notation is a special format to write extremely big or small numbers in shorthand, but you won't only see it in science class. Scientific notation has probably popped up on your calculator before, and you will see it is used in media, finance, demographics, and yes, science. Scientific notation makes it easier to compare and utilize very large and small numbers as you'll see throughout this lesson.

Why Scientific Notation?

Science, finance, and population are just some areas that involve incredibly large or small numbers. Scientists writing papers about outer space don't want to have to keep writing huge numbers like 93,000,000 miles (the distance from the sun to the Earth) and 300,000,000 (the approximate speed of light in meters per second), so scientific notation gives us a universally accepted way to discuss large and small numbers. We can also apply the properties of operations and the properties of exponents to numbers written in scientific notation. This allows us to easily perform calculations on numbers that would otherwise be very cumbersome to work with.

Multiplying and Dividing by Powers of 10

Scientific notation was developed on the foundation that our number system is based on powers of ten. Before we get into the meat of scientific notation lets take a moment to review multiplying by powers of 10. The powers of 10 include 1, 10, 100, 1,000, and so on. Some examples of the powers of ten less than 1 include 0.1, 0.01, and 0.001. To multiply by a power of 10 greater than one, simply move the decimal place once to the right for every zero in the power of 10: For example, $7.84 \times 1{,}000 = 7{,}840$ and $0.003 \times 10{,}000 = 30$. Multiplying by a power of 10 that has a value that is less than 1 will make a number smaller. To multiply by a power of 10 less than 1, simply move the decimal place once to the left for every place value in the power of 10: For example, $430 \times 0.1 = 43$ and $15 \times 0.001 = 0.015$.

General Format of Scientific Notation

What does it mean to write a number in scientific notation? The general format for a number written in scientific notation is $a \times 10^b$, with specific requirements for the types of numbers a and b can be:

1. a must always be a number greater than zero and less than 10.
2. b is a positive or negative integer that determines the number of spaces the decimal point moves to the left or to the right (b can also be 0).

Scientific notation represents numbers in the format: $a \times 10^b$, where $0 < a < 10$ and b is an integer. When $b \geq 0$ the number will be greater than or equal to one. Conversely, when $b < 0$ the number will be less than one.

Converting from Scientific Notation to Standard Notation

Converting a number from scientific notation to standard notation is fairly straightforward since the exponent gives clear instructions on how many times to move the decimal right or left. For example, the scientifically notated number 5.9×10^4 translates to 59,000. Since 10^4 has four zeros (10,000), the decimal in 5.9 moves to the right four times. Negative exponents will move the decimal the other direction. Why? You learned in the previous lesson that a negative exponent moves the base into the denominator of a fraction. Therefore, $10^{-1} = \frac{1}{10^1} = 0.1$ and $10^{-2} = \frac{1}{10^2} = 0.01$. When a number is multiplied by a power of 10 with a negative exponent, the decimal moves to the left the same number of times as the exponent. Using this trick we can see that the expression 8.5×10^{-3} represents the number 0.0085.

Converting from Standard Notation to Scientific Notation

Converting a number from standard notation to scientific notation is generally a little trickier for students. Let's convert 9,800,000 in scientific notation. Since 9,800,000 must be written in the format $a \times 10^b$, where a is between zero and 10, our decimal must go between the 9 and the 8, making $a = 9.8$. Since this placement moved the *original* decimal six places to the left, we will write the number as 9.8×10^6. We can check our answer by sliding the decimal point in 9.8 six times to the *right*; since we see that gives us our original value of 9,800,000, we know we have the correct scientific notation.

Suppose we have a really small number that needs to be written in scientific notation, like 0.000047. We are going to approach this task the same way. First, we need to select an a that is between zero and 10, so we put the decimal between the 4 and 7 to get 4.7. Next, we need to multiply by a power of 10 to make the scientific notation represent the original number. To get to 4.7, we moved the decimal point five spaces to the *right*. We reflect this in the power of 10 by using a negative exponent: 4.7×10^{-5}. We can verify that this answer is correct by sliding the decimal point in 4.7 five times to the *left*; since that gives us our original value 0.000047, we know that 4.7×10^{-5} is the correct scientific notation.

Practice 1

1. Which of the following choices are correctly displayed in scientific notation?
   a. 0.8×10^3
   b. 10×10^{-1}
   c. 7.0003×10^2
   d. 5×0.1^{10}

2. How is the number 10 written in scientific notation?

3. When a number less than one is expressed in the scientific notation format $a \times 10^b$, what is true about the value of b?

4. Translate 316.72 into scientific notation.

5. Express 0.00205 in scientific notation.

6. Pluto is 5,914,000,000 km from the sun. Represent this distance in scientific notation.

7. The width of a specialized medical probe is 0.0008 centimeters. Represent this width in scientific notation.

8. Mothers Against Drunk Driving (MADD) estimates that approximately 3.3×10^4 people died in 2014 due to drunk driving. What is this number in standard form?

9. If the average person blinks 6.24×10^6 times a year, express this number in standard form.

10. The diameter of a grain of sand is approximately 2.4×10^{-3} inches. Express its diameter in standard form.

Performing Operations with Scientific Notation

One benefit of writing extremely large or small numbers in scientific notation is that the properties of operations and the laws of exponents can be applied to them in order to multiply and divide numbers in scientific notation.

Dividing Numbers in Scientific Notation

To perform division between two numbers in scientific notation form $a \times 10^b$, divide the a terms and use the rules of exponents to simplify the 10^b terms.

Example: $\frac{7 \times 10^9}{2 \times 10^5}$

Solution:

Divide the a terms: $\frac{7}{2} = 3.5$

Use the laws of exponents to simplify the 10^b terms:

$\frac{10^9}{10^5} = 10^{(9-5)} = 10^4$

Therefore, $\frac{7 \times 10^9}{2 \times 10^5} = 3.5 \times 10^4$

ERROR ALERT! It is easy to make a mistake when performing division between negative exponents. Remember to use *keep-switch-switch* to rewrite any subtraction as addition: $\frac{8 \times 10^2}{2 \times 10^{-4}} = 4 \times 10^{[2-(-4)]} = 10^{(2+4)} = 10^6$

Multiplying Numbers in Scientific Notation

To perform multiplication between two numbers in scientific notation form $a \times 10^b$, multiply the a terms and use the rules of exponents to simplify the 10^b terms.

Example: $(7.2 \times 10^{-2}) \times (3 \times 10^5)$

Solution:

Multiply the a terms: $7.2 \times 3 = 21.6$

Use the laws of exponents to simplify the 10^b terms: $10^{-2} \times 10^5 = 10^{(-2+5)} = 10^3$

So $(7.2 \times 10^{-2}) \times (3 \times 10^5) = 21.6 \times 10^3$, but we are not done yet. Since the a term is not between 0 and 10 we must take one power of 10 *away* from the a term by moving the decimal to the left, and give that power of 10 to the 10^b term by increasing b by 1. Rewrite 21.6×10^3 as 2.16×10^4, which is the final answer.

ERROR ALERT! Many students find it difficult to rewrite terms that are partially in scientific notation, like 21.6×10^3, as numbers that are in complete scientific notation, like 2.16×10^4. In order to do this successfully, make sure if you are making the *a* term smaller and you are making the 10 term bigger by adding one to the *b*. This way you will create an equivalent numerical value. (Conversely, when you are dividing numbers in scientific notation, you might end up with an *a* term that is less than one. In that case, you will need to make your *a* term bigger and then make the 10^b term smaller by subtracting one from the *b*.)

Practice 2

1. Calculate and write the answer in scientific notation: $\frac{9 \times 10^6}{4.5 \times 10^{-2}}$

2. Calculate and write the answer in scientific notation: $(5.3 \times 10^{-2}) \times (8 \times 10^5)$

3. What is 5×10^{-2} divided by 4×10^{-7}?

4. Simplify $(2.2 \times 10^3) \times (3 \times 10^{-5})$

5. A family trust worth 1.6×10^7 is being divided between 8 family members. Use the shortcuts for dividing numbers in scientific notation to find how much money each family member will receive. Write your answer in scientific notation and in standard form.

6. The San Francisco Bay Area includes Silcon Valley and is one of the tech capitals of the world. The Bay Area has about 3.9×10^5 tech jobs and the average tech salary pays $\$1.44 \times 10^5$ per year. Use the shortcuts for multiplying numbers in scientific notation to find out the approximate amount of money that is earned each year in the Bay Area just in the technology sector. Write your answer in scientific notation, in standard form, and in expanded form (write out the number in words).

7. Write out each of the two numbers from question 6 in standard notation:

3.9×10^5 = ________________

$\$1.44 \times 10^5$ = ________________

Now multiply these two numbers in standard notation and check your answer with your response for question 6. Did you get the same answer? (If not you have an error.) Which method of multiplication was easier to perform? Is multiplying in scientific notation more or less efficient than multiplying in standard form?

Scientific Notation All Around Us

So, how does scientific notation play out in the real world? Suppose we were measuring a planet's distance from the sun in kilometers. That would end up being a really large number! Instead of writing a long number with many zeros, we can simply and concisely write it using scientific notation. Or suppose we need to write the mass of a single particle of dust, which is an absurdly small number. Instead of writing a long decimal with a value of less than one, we can concisely write it in scientific notation.

Quite often when we are working with numbers in scientific notation, we are going to be dividing them to calculate a rate or we will be multiplying them to get a bigger picture of a trend. Remember to always check to see if your answer makes real-world sense. If you are calculating how many more times people live in India than in New York City and you get two, you probably made a mistake somewhere! Let's do one example together before you practice some on your own.

PROBLEM-SOLVING TIP

Solving real-world problems that are represented in scientific notation can be difficult because their appearance can distract us from our mathematical intuition. In order to determine if a problem requires multiplication or division, replace the scientific notation numbers with friendly whole numbers and then work to determine which operation would correctly solve the problem. We will show how this technique is used in the following problem.

Example: The total combined earnings of a country in the developing world is 8.4×10^{11} dollars. If the country has 4.2×10^{7} households, calculate the average income per household.
Solution: Let's temporarily replace the scientific notation numbers with other potential numbers. If a country had a combined earnings of $10,000 and there were 20 families, we would divide $10,000 by 20 to determine the average income. This helps us to see that we need to use division with the scientific notation numbers in this problem.

Write the division as a fraction: $\frac{8.4 \times 10^{11}}{4.2 \times 10^{7}}$

Divide the a terms: $\frac{8.4}{4.2} = 2$

Use the laws of exponents to simplify the 10^{b} terms: $\frac{10^{11}}{10^{7}} = 10^{(11-7)} = 10^{4}$

Therefore, $\frac{8.4 \times 10^{11}}{4.2 \times 10^{7}} = 2 \times 10^{4}$

The average income per household represented by 2×10^{4} is $20,000.

Practice 3

1. Determine how many seconds are in a day and represent this number in scientific notation.

2. Milli's vet told her that, on average, an active dog burns 8.1×10^{-3} calories per second. Multiply your answer from question 1 by 8.1×10^{-3} to see approximately how many calories an active dog burns in a day.

Use the following information to answer questions 3 and 4:

A country receives 30 million dollars in aid after a natural disaster. There are forty thousand families that need aid money to help them recover from the disaster.

3. How would 30 million and 40 thousand be represented in scientific notation?

4. What operation would need to be performed to solve this problem? Determine how much aid money each family will get.

5. The population of India is around 1,252,000,000 and the population of New York City is approximately eight and a half million people. Write an expression using scientific notation that could be used to determine how many times more people live in India than in New York City.

6. Solve your expression from question 5.

Answers

Practice 1

1. Choice **c** is the only number represented in scientific notation: 7.0003×10^2
2. $10 = 1 \times 10^1$
3. When a number less than one is expressed in the scientific notation format $a \times 10^b$, b will be less than zero.
4. $316.72 = 3.1672 \times 10^2$
5. $0.00205 = 2.05 \times 10^{-3}$
6. $5{,}914{,}000{,}000 = 5.914 \times 10^9$
7. $0.0008 = 8 \times 10^{-4}$
8. $3.3 \times 10^4 = 33{,}000$
9. $6.24 \times 10^6 = 6{,}240{,}000$
10. $2.4 \times 10^{-3} = 0.0024$

Practice 2

1. 2×10^8
2. 4.24×10^4
3. 1.25×10^5
4. 6.6×10^{-2}
5. $1.6 \times 10^7 \div 8 = \frac{1.6 \times 10^7}{8 \times 10^0} = 0.2 \times 10^7 = 2 \times 10^6 = \$2{,}000{,}000$
6. $(3.9 \times 10^5) \times (\$1.44 \times 10^5) = 5.616 \times 10^{10} = \$56{,}160{,}000{,}000$. Each year 56 billion, 160 million dollars is earned in the technology sector in the Bay Area.
7. You should get the same answer. It is likely that you prefer multiplying numbers in scientific notation (unless you are a glutton for punishment and absolutely love multiplying huge numbers).

Practice 3

1. Seconds per day $= 60 \times 60 \times 24 = 86{,}400 = 8.64 \times 10^4$
2. $(8.64 \times 10^4) \times (8.1 \times 10^{-3}) = 69.984 \times 10^1 = 6.9984 \times 10^2$
3. *30 million* $= 3 \times 10^7$ and *forty thousand* $= 4 \times 10^4$
4. Divide the total amount of money received by the number of people in need: $(3 \times 10^7) \div (4 \times 10^4) = 0.75 \times 10^3 = 7.5 \times 10^2 = 750$. This \$30 million will only give \$750 to each of the 40,000 people in need.

5. $1{,}252{,}000{,}000 = 1.252 \times 10^9$ and eight and a half million people $= 8.5 \times 10^6$. In order to see how many times more people live in India than in New York City, we would divide the number of people in India by the number of people in New York City: $\frac{1.252 \times 10^9}{8.5 \times 10^6}$

6. $\frac{1.252 \times 10^9}{8.5 \times 10^6} = 0.1473 \times 10^3 = 1.473 \times 10^2 = 147.3$. The population in India is more than 147 times the population of New York City.

21

Understanding Slope

STANDARD PREVIEW

In this lesson we will cover Standard 8.EE.B.5. You will learn that the slope of a line is a numerical and visual representation of the rate of change between the dependent and independent variables. You will also learn how to calculate the slope and how to discuss it in real-world terms.

Comparing Different Representations of Relationships

Sometimes you'll get an invitation to a birthday party as an email. Sometimes your invitation might come verbally, by word of mouth. Other times you might actually get a paper invite in the mail! Similarly to how there are different ways to learn about the same birthday party, there are different ways to receive information about the same mathematical relationship between two factors, like *time* and *pay*. We can receive mathematical information as an algebraic equation, as entries in a table, or in a verbal description of coordinating pieces of information. Regardless of *how* we receive mathematical information, we can compare these various forms of information by plotting the mathematical relationships they represent as lines in a coordinate axis graph. Graphs are useful tools for comparing data because the *slope* or steepness of a line tells us important information about the rate of change. Let's learn how to compare different forms of information by using graphs and how to interpret the slopes of lines. We'll begin with some students who are discussing the money they earn for chores at home.

Plotting Information on a Graph

Jonah and Auggie talk about what their parents pay them for certain chores. Auggie's parents pay him \$12 for every hour and a half of leaf raking he completes. Jonah's parents use an equation of $\$8h = p$ to determine his pay for mowing the lawn. His parents plug in the number of hours Jonah works for h to calculate his pay, p. Jonah and Auggie want to compare the two different ways their parents pay them to see who earns more money. They agree that Jonah can use the equation $\$8h = p$ to make a table showing the relationship between hours and pay. Auggie will make a similar table using proportional relationships. Then they will graph the ordered pairs in each of their tables on two different graphs to show their pay for up to six hours of work.

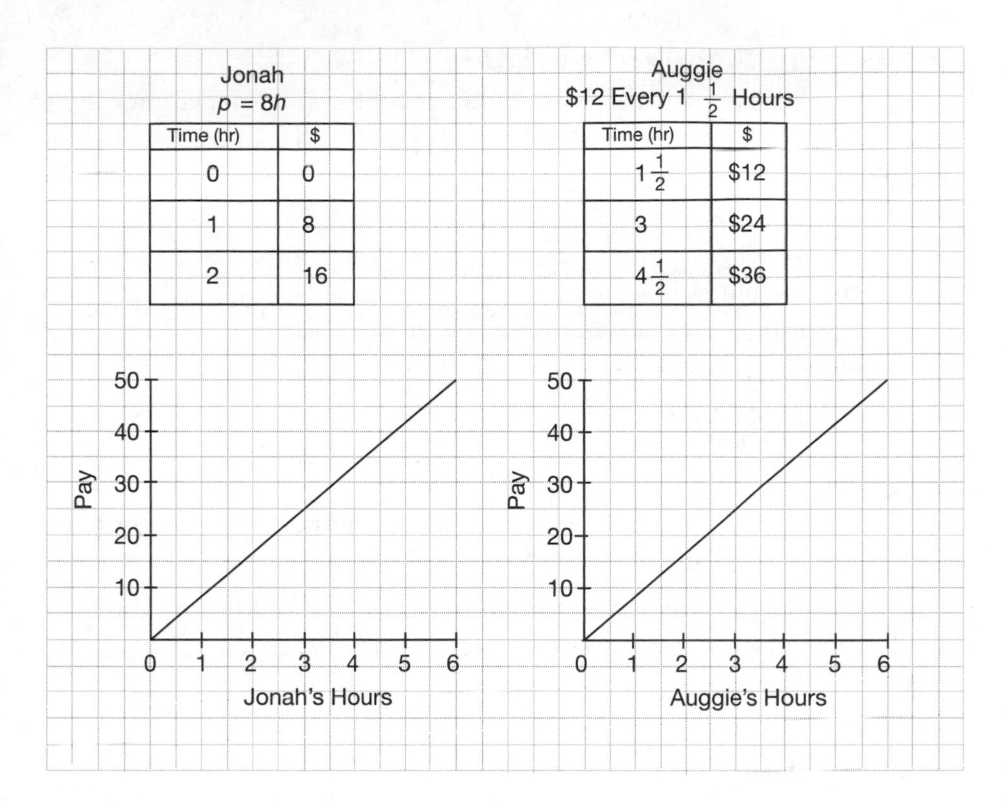

Time (hr)	$
0	0
1	8
2	16

Time (hr)	$
$1\frac{1}{2}$	$12
3	$24
$4\frac{1}{2}$	$36

Jonah and Auggie notice that their graphs look identical and their lines have the same exact steepness. This means the relationship between their independent variable, *time*, and the dependent variable, *pay*, is the same. Auggie looks closely at his graph and sees that when his *time* goes from two hours to three hours, his *pay* goes from $16 to $24, so he determines that his pay rate is $8 per hour. (He's a little disappointed because it sounded like he was earning more than Jonah since he got $12 for an hour and a half of work.) How was Jonah able to see that his pay was $8 for each hour, just by looking at the equation his parents use, $p = 8h$?

Slope as Rate of Change

Auggie and Jonah have just determined that both of their pay is changing at a rate of $8 per hour. This brings us to a special type of rate: slope. **Slope is the "rate of change" that compares the change in the dependent variable to the change in independent variable.** Therefore, since *pay* is dependent and *time* is independent in this relationship, the slope of Auggie and Jonah's pay is calculated as:

$$\text{Slope} = \frac{\text{change in pay}}{\text{change in hours}}$$

Calculating Slope from a Table

Stella, who babysits her baby sister, joins the boys' conversation. She's recorded her past couple of payments in this table. (She spilled some juice on the table, so the last entry is illegible.)

Minutes	Pay
15	$2.50
45	$7.50
75 minutes	$12.50
90 minutes	(juice smudge)

The boys know that the slope, or rate of change, of their pay is $8 per hour and they decide to calculate Stella's pay slope. They will use the ordered pairs from Stella's table, (75 minutes, $12.50) and (15 minutes, $2.50) to calculate the *change in pay* divided by the *change in time*:

$$\text{Slope} = \frac{\text{change in pay}}{\text{change in hours}} = \frac{\$12.50 - \$2.50}{75\text{ min} - 15\text{ min}} = \frac{\$10}{60\text{ min}}$$

When done, they convert 60 minutes into 1 hour and write the ratio as $\frac{\$10}{1}$. This shows that Stella is making $10 for every 1 hour she works.

Comparing Different Slopes on a Graph

The boys are feeling a little envious and want to see what Stella's rate of $10/hour looks like on a graph, so they plot her pay-scale line next to theirs. Since she makes $10 an hour, they plot points (1 hr,$10) and (3 hr,$30). They represent her pay with the dashed line:

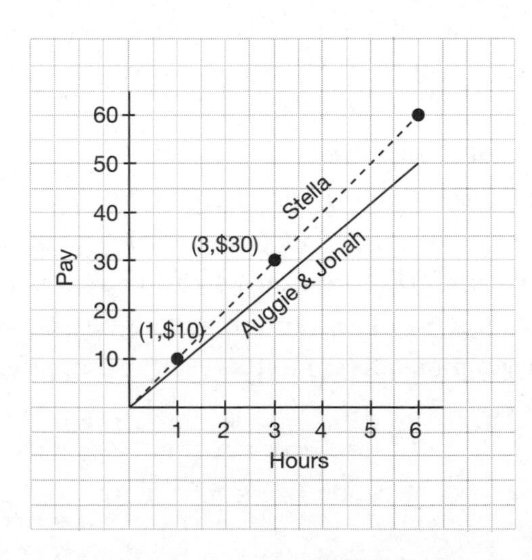

Notice that the line modeling Stella's pay is *steeper* than the line modeling Auggie and Jonah's pay. This is because the slope of Stella's income is 10 and the slope of the boys' income is 8. **The steepness of a line is an indicator of its rate of change, or slope. The greater a line's slope is, the steeper it will be when graphed.**

Finding the Slope between Two Points

Remy's parents pay for her to have data on her cell phone each month, so she doesn't earn too much money for helping out with chores around the house. Last weekend she did $2\frac{1}{2}$ hours of work and they gave her $10. Yesterday her mom gave her $18 after she helped paint the kitchen Splash Blue for $4\frac{1}{2}$ hours. Remy wants to plot the points $(2\frac{1}{2},10)$ and $(18,4\frac{1}{2})$ on the graph, but luckily Stella realizes that the second point $(18,4\frac{1}{2})$ needs to be switched to $(4\frac{1}{2},18)$ so that the *independent* variable of *time* comes before the *dependent* variable of *pay* in both points. The students add $(2\frac{1}{2},10)$ and $(4\frac{1}{2},18)$ to their graph and connect them with a dotted line:

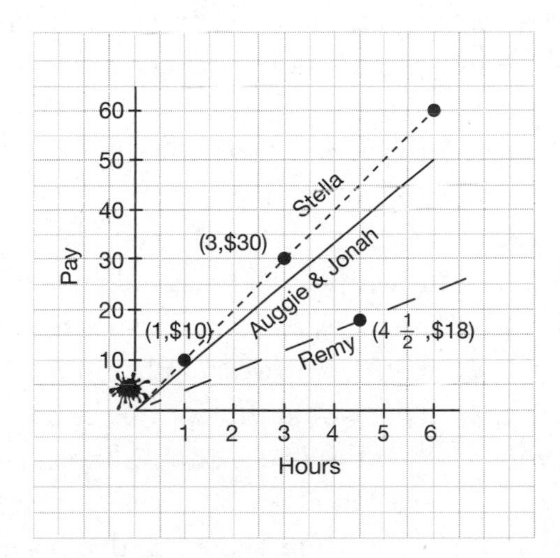

The students notice that the slope of Remy's hourly pay line is less steep than the other two lines. Therefore, Remy must be making less per hour than they are. **The more gradual a line is, the smaller its slope is.** They try to read how much she makes *per hour* by looking at the *pay* on the *y*-axis when *hours* = 1 on the *x*-axis, but by now there's a juice splash on the corner of the graph, so they can't be certain what her pay for 1 hour is. They decide to use the slope formula for the points ($2\frac{1}{2}$ hr, \$10) and ($4\frac{1}{2}$ hr, \$18) to calculate the rate at which Remy's pay changes (her *pay per hour*):

$$\text{Slope} = \frac{\text{change in pay}}{\text{change in time}} = \frac{\$18 - \$10}{4\frac{1}{2}\text{ hr} - 2\frac{1}{2}\text{ hr}} = \frac{\$8}{2\text{ hr}}$$

They realize that an extra \$8 pay for 2 hours of time is the same thing as getting paid \$4 per hour. They reduce $\frac{8}{2}$ to $\frac{4}{1}$, but keep it as a fraction since that shows the relationship between *pay* and *time*. Jonah, Auggie, and Stella make fun of Remy for earning less than they do per hour, but she just shrugs it off and smiles, knowing that none of them have cell phones with data!

Practice 1

1. What is slope?

2. What does it mean when the slope between two points is steep? What does it mean when the slope between two points is gradual?

3. Does it show more information to write a slope of seven as 7 or as $\frac{7}{1}$? Explain your reasoning.

4. If graphed on the same axis, which line would likely have a steeper slope: line T, which compares *number of months* to *height* for a 12 year old or a line F, which compares number of months to height for an 18-year-old? What would the slope be of line S that compares *number of months to height* for a 50 year old? Explain your reasoning.

5. Another student joins the group, reporting that his uncle paid him \$60 for eight hours of work he did last month and \$15 for 2 hours of work he did last weekend. Express this information as two coordinate pairs, find the slope between these two points, and reduce the fraction to lowest terms. What does this ratio represent in terms of hours and pay?

6. You should have kept your answer in question 5 in lowest terms, which means the denominator equaled two. Now, write this as a *unit rate* by making an equivalent fraction with a denominator of one. What does this new ratio represent in terms of hours and pay?

The Nuts and Bolts of Slope

It's important to remember that the line that represents a relationship is *proportional* if it goes through the point (0,0). As we will see in the next lesson, there are many cases where a line does not go through the origin. Up until now, this lesson has presented situations where if the students worked for zero hours, they got zero dollars, so all of these relationships were proportional. The slope in all proportional relationships can be found by dividing any y-coordinate by its corresponding x-coordinate. We have instead been practicing finding the slope by using the rate of change because that is the method that must be used in all non-proportional relationships. There is a formula to calculate slope that can be used for *all* coordinate pairs, regardless of whether they are proportional or not. In this section we are going to work with this formula and learn how we can discuss slope as a mathematical relationship in words. Let's get started!

The Slope Formula

There is a formula used to find the slope between *any* two ordered pairs, regardless of whether or not they are proportional. We will write the pairs (x_1,y_1) and (x_2,y_2). (This subscript notation of the tiny 1s and 2s is used often in algebra. It is not a weird mathematical operation, but rather just a method for indicating which x-coordinate or y-coordinate is being referred to.) All over the world, the variable used to represent slope is m. Why? We have no idea, but if you remember that m in algebraic equations is the slope, it'll make your math life easier! Here's the conventional slope formula:

$$\text{Slope} = m = \frac{y_2 - y_1}{x_2 - x_1}$$

ERROR ALERT! This formula is used to find the slope between any two points (x_1,y_1) and (x_2,y_2). It is all too common for students to subtract the x-coordinates in the numerator and to put the y-coordinates in the denominator. We guarantee that you will make this easy mistake at least once, but if you know to look out for it and be careful, perhaps you'll make it *just* once!

Example: Find the slope between the two points (–2,–42) and (8,–12)
Solution:

Step 1: Label x_1, y_1, x_2, and y_2 under the points:

$$\underset{x_1}{(-2},\underset{y_1}{-42)} \text{ and } \underset{x_2}{(8},\underset{y_2}{-12)}$$

Step 2: Write the formula: **Slope** = $m = \frac{y_2 - y_1}{x_2 - x_1}$

Step 3: Plug in values carefully and evaluate:

$$\textbf{Slope} = m = \frac{y_2 - y_1}{x_2 - x_1} = \frac{-12 - (-42)}{8 - (-2)} = \frac{-12 + 42}{8 + 2} = \frac{30}{10}$$

Step 4: Reduce the slope.

Slope = $\frac{30}{10} = \frac{3}{1}$, so for every 1 change in x, y will change 3.

ERROR ALERT! Did you notice how the coordinate labels, x_1, y_1, x_2, and y_2 are written under the coordinate pairs? Do this! While it is *not* important which y-coordinate is first in the numerator, it is *critical* that the *corresponding x-coordinate* go first in the denominator. A very common error is to use the y-coordinate from one point first in the numerator and then the *non-corresponding x*-coordinate first in the denominator. This will be especially tempting when the coordinate pairs include negative values! Labeling your pairs in this manner will help you input numbers correctly into the slope formula!

Slope is the rate of change between two different quantities. It is the change of the dependent variables (y) divided by the rate of change of the independent variables (x). Slope can be calculated by using any two points (x_1,y_1) and (x_2,y_2), taken from a graph, from a table, or from a word problem. *m* is the variable that represents slope:

$$\textbf{Slope} = m = \frac{\text{change in } y}{\text{change in } x} = \frac{y_2 - y_1}{x_2 - x_1}$$

Intelligently Interpreting Slope

Jackson joins the group of aforementioned students and reports that he worked 22 hours during the first week of spring break and earned $209. During the second week of break he earned $266 for 28 hours of work. Since these points do not fit on the previous graph, the students are not certain if these two points are proportional and will create a graph that goes through the point (0,0). They decide to use the slope formula to find the rate of change of Jackson's pay. They write the given information as points (22 hours,$209) and (28 hours,$266) and reduce the slope to simplest terms after calculating it:

$$\underset{x_1}{(22 \text{ hours}},\underset{y_1}{\$209)} \text{ and } \underset{x_2}{(28 \text{ hours}},\underset{y_2}{\$266)}$$

$$\textbf{Slope} = \boldsymbol{m} = \frac{y_2 - y_1}{x_2 - x_1} = \frac{\$266 - \$209}{28 \text{ hours} - 2 \text{ hours}} = \frac{\$57}{6 \text{ hours}}\left(\frac{\div 3}{\div 3}\right) = \frac{\$19}{2 \text{ hours}}$$

Remember *intelligent rounding* from a few lessons back? It was where you had to apply the real-life context to an answer to decide if it made sense to round a decimal up or down, rather than just blindly following the rounding rules you learned in elementary school. It is important to be able to also perform *intelligent interpretation* of slope, which means being able to clearly discuss it as a mathematical relationship in words. Let's consider Jackson's pay slope of $\frac{19}{2}$. To say that Jackson's rate of pay is *19 over two* doesn't make any real-world sense, but to say that *for every two hours Jackson works, he earns $19*, now *that* is intelligent! When explaining a slope relationship in words, it is critical to make sure that you are demonstrating that the value in the denominator is determining the value in the numerator. If there is a negative number in your numerator or denominator, then that means that the y-variable is *decreasing* as the x-value is *increasing*, which is critical to make clear. (We do not usually give the negative value to the x-variable because it makes more sense to always talk about what y is doing when x is *increasing*.) Here are some examples:

> $m = \frac{5}{3}$: This slope could represent that for every three days, a baby is gaining five ounces.
> $m = -\frac{1}{7}$: This slope could represent that for every seven days a person is on a diet, he or she is losing one pound.
> $m = \frac{55}{1}$: This slope could represent that for every one hour a car is driving, it covers 55 miles.
> $m = -\frac{1}{3}$: This slope could represent that for every one mile a couple is hiking, the distance to their final destination is reducing by three miles.

It is important to notice that with both of the negative slopes, the attribute of "decreasing" was given to the dependent variable in the numerator and not to the independent variable in the denominator.

Slope as Unit Rate in Proportional Relationships

Although it is standard practice to keep slope in simplest terms without reducing the denominator to one, if you are certain that a relationship is proportional, that it will contain the point (0,0), then reducing the denominator to one has a benefit: if we write $\frac{19}{2}$ as the equivalent ratio $\frac{9.5}{1}$, we see that Jackson earns $9.50 per 1 hour. **In proportional relationships, the slope reduces to the unit rate.** Knowing this, we don't need to graph his income to see that he earned more than Jonah, Auggie, and Remy, per hour, but not as much as Stella. What would this mean about the steepness of the line that models Jackson's pay? The slope of his pay line would be steeper than three of the lines on the graph, but not as steep as Stella's line.

Practice 2

1. What are the two most common errors students make when calculating slope?

2. The slope of a line is eight. What does that mean? Write it as a ratio and make up a scenario this slope could represent.

Use the following information to answer questions 3 through 5:

On January 1st of 2009, the population of Tinytown was 4,984 people. On January 1st of 2015, the population of Tinytown was 4,878.

3. Write two coordinate pairs that correctly represent the independent and dependent variables regarding the population in Tinytown.

4. Calculate the slope and describe in words what the slope in simplest terms is representing.

5. Calculate the unit rate for the population growth/decline in Tinytown and describe in words what this unit rate represents.

6. Find the slope between the two coordinate pairs and create a story to go along with it: (8,–1) and (6,3).

7. Find the slope between the two coordinate pairs and create a story to go along with it: (9,3) and (–1,7).

Answers

Practice 1

1. Slope is the rate of change that compares the change in the dependent variable to the change in independent variable.
   Slope = $\frac{\text{change in } y\text{-coordinates}}{\text{change in } x\text{-coordinates}}$
2. When the slope is steep a relationship has a fast or high rate of change. When the slope is gradual, it means the slope is a smaller number.
3. It shows more information to write a slope of seven as $\frac{7}{1}$ because this shows the relationship between the dependent and independent variables. Every one change in the independent variable makes the dependent variable change seven.
4. Since 12 year olds are growing at a faster rate than 18 year olds, the slope of line T would be steeper than the slope of line F. Since someone in her 50s isn't growing anymore, her slope would be a flat line to show that her height is not getting bigger or smaller.
5. (2,$15) and (8,$60)
   Slope = $\frac{\text{change in } y\text{-coordinates}}{\text{change in } x\text{-coordinates}} = \frac{\$60 - \$15}{8 - 2} = \frac{\$45}{6} = \frac{\$15}{2 \text{ hours}}$
   This boy earned $15 for every 2 hours of work.
6. $\frac{\$15}{2 \text{ hours}}$ is the same thing as $\frac{\$7.50}{1 \text{ hour}}$, which means he earns $7.50 per hour from his uncle.

Practice 2

1. When calculating slope students often make these mistakes: 1) put the x-coordinates in the numerator instead of the denominator; 2) use the y-coordinate from one point first in the numerator and then the *non-corresponding* x-coordinate first in the denominator; 3) do not use *keep-switch-switch* to write confusing subtraction as addition and make mistakes with the arithmetic.
2. The slope of a line is $8 = \frac{8}{1}$. This could mean that for every one hour Rio worked, he created eight invoices for his boss's clients.
3. (2009, 4,984) (2015, 4,878)
4. **Slope** = $\boldsymbol{m} = \frac{y_2 - y_1}{x_2 - x_1} = \frac{\text{people}}{\text{years}} = \frac{(4{,}878 - 4{,}984)}{(2015 - 2009)} = \frac{-106}{6}\left(\frac{\div 2}{\div 2}\right) = \frac{-53}{3}$
   For every three years, the population of Tinytown decreased by 53 people.
5. The unit rate is the slope reduced to a denominator of one: $\frac{-53}{3} = \frac{-17.6}{1}$. This shows that every year, the population decreased on average around 17 to 18 people.

6. **Slope** = $m = \frac{y_2 - y_1}{x_2 - x_1} = \frac{(3 - -1)}{(6 - 8)} = \frac{4}{-2} = \frac{-2}{1}$

For every minute that passed, the sun dropped two more degrees toward the horizon. (Make sure to gave the negative characteristic to your dependent variable and not your independent variable.)

7. **Slope** = $m = \frac{y_2 - y_1}{x_2 - x_1} = \frac{7 - 3}{-1 - 9} = \frac{4}{-10} = \frac{-2}{5}$

The traffic was heavy, but not terrible, so for every five minutes Thelma and Louise drove on the freeway, their remaining distance to travel decreased by two miles. (Make sure to give the negative characteristic to your dependent variable and not your independent variable.)

22

Writing Slope-Intercept Linear Equations

STANDARD PREVIEW

In this lesson we will cover **Standard 8.EE.B.6**. You will learn that "rise over run" is another tool to calculate slope and will learn the difference between lines in $y = mx$ form and $y = mx + b$ form. You will also learn what the y-intercept is and how to write the slope-intercept form of lines given information presented in different forms.

Slope as "Rise Over Run"

We learned lots of things about slope in the previous lesson: slope is a rate of change, we can determine the slope from information presented in lots of different ways, how slope can be described in real-world context, and the formula for calculating the slope between any two points. Now you are going to learn how the jingle "rise over run" is another useful slope tool. This little slogan is taught to help students remember that slope is the *change in the y-coordinates* (or the *rise*) divided by the *change in the x-coordinates* (or the *run*). The word *rise* represents the vertical change of a line, as compared to the corresponding *run*, or horizontal change of a line. Some students like to use $\frac{\text{rise}}{\text{run}}$ when they are writing out their slope formula:

$$\textbf{Slope} = m = \frac{\text{rise}}{\text{run}} = \frac{y_2 - y_1}{x_2 - x_1}$$

Rise Over Run Slope Triangles

Emery Carey is buying sheets of stickers for Anna Hale at their favorite toy store. If every three sheets of stickers cost $2, we can graph this relationship by plotting the points (three sheets, $2) and (six sheets, $4) and connecting them. We can create a **slope triangle** between these two points to illustrate that the *rise*, or *vertical change in price*, is $3, while the *run*, or *horizontal change in number of sheets*, is two. Notice that the dotted triangle below provides a mapping that illustrates the relationship between points (2,3) and (4,6).

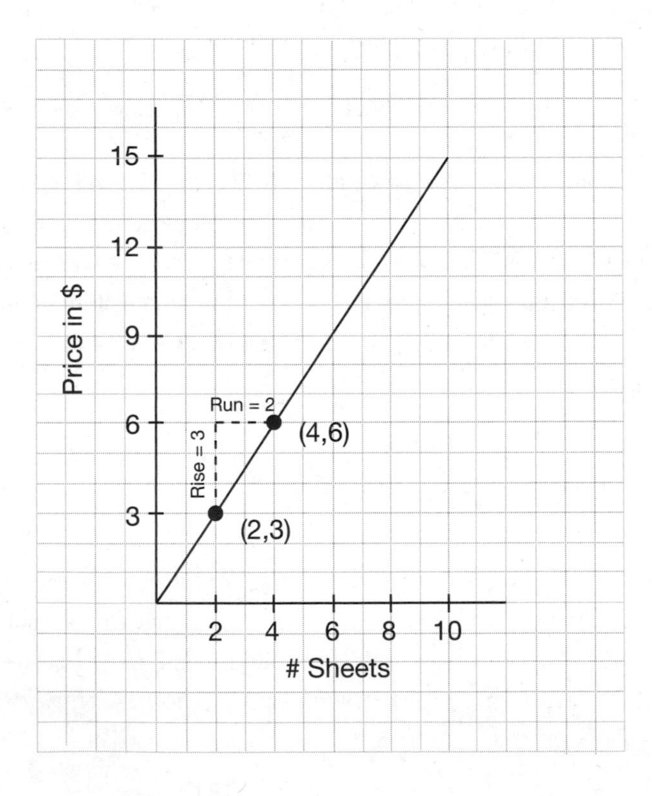

Now that we've done a rise over run slope triangle we can plug our rise and run into the slope formula to calculate the slope:

$$\textbf{Slope} = m = \frac{\text{rise}}{\text{run}} = \frac{3}{2}$$

The previous slope triangle shows us that the slope of this line is $\frac{3}{2}$. Let's look at how our slope changes if we create an entirely different slope triangle between two different points on the same line. The following shows the addition of a slope triangle between the points (6,9) and (10,15):

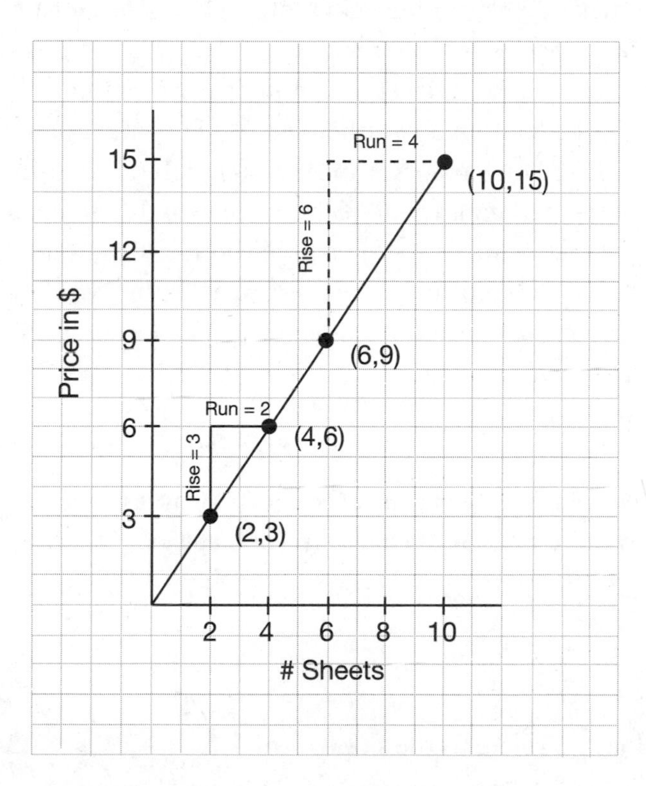

The new slope triangle shows that a *rise* of six corresponds with a *run* of four, so we'll put those numbers into our $\frac{\text{rise}}{\text{run}}$ slope formula:

$$\textbf{Slope} = m = \frac{\text{rise}}{\text{run}} = \frac{6}{4}\left(\frac{\div 2}{\div 2}\right) = \frac{3}{2}$$

It is not a coincidence that the second slope triangle has a $\frac{\text{rise}}{\text{run}}$ that reduces to $\frac{3}{2}$. Can you identify anything special about the proportions of these two slope triangles?

Slope between All Collinear Points

Did you recognize that the previous slope triangles are *similar* triangles and have the same proportions? Similar triangles have proportional sides and all slope triangles belonging to the same line are *similar*. Therefore, the ratios between the corresponding sides of any two slope triangles on the same line are equivalent. This means that the *rise over run* (slope) created by one slope triangle will be the same as the *rise over run* (slope) of a different slope triangle mapping two different points on the same line. Since *rise over run* is the same as slope, and since points on the same line are called *collinear*, we make the very important conclusion: ***The slope between any pairs of collinear points is the same.***

This is big news because it lets us know that we can find the slope of a line by using any two points in either the $\frac{y_2 - y_1}{x_2 - x_1}$ formula or a *rise over run* slope triangle. If you pick points A and B and find the slope using a rise over run slope triangle, and your friend picks points C and D and determines the slope by using the formula $\frac{y_2 - y_1}{x_2 - x_1}$, you will both get the same exact slope after it has been reduced to simplest terms! Whether you are looking at the beginning, middle, or end of a line, the slope of a straight line remains constant and does not change.

The slope between any distinct pairs of collinear points is the same. The slope of a line remains constant throughout any part of that line.

Practice 1

1. How does $\frac{\text{rise}}{\text{run}}$ relate to the slope formula, $\frac{y_2 - y_1}{x_2 - x_1}$?
2. What is true about all slope triangles drawn on the same line?
3. If two different pairs of points are collinear, what can be said about the slopes between both pairs of points?

Use the following graph to answer questions 4 through 7:

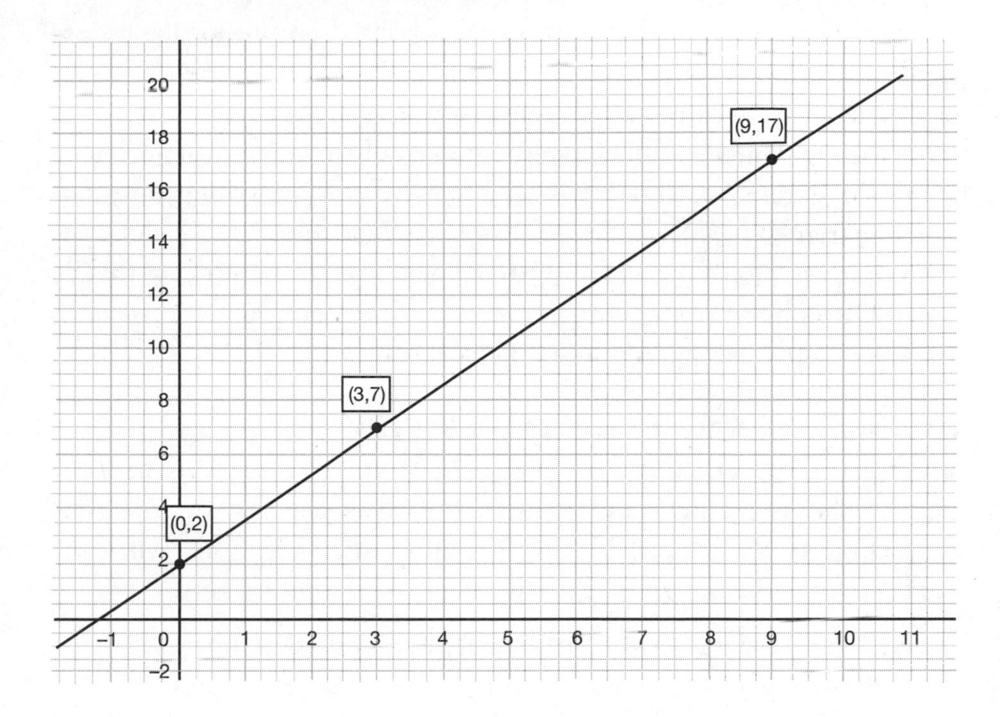

4. Make a slope triangle between the points (0,2) and (3,7). Determine the $\frac{\text{rise}}{\text{run}}$ and reduce to lowest terms.

5. Make a slope triangle between the points (3,7) and (9,17). Determine the $\frac{\text{rise}}{\text{run}}$ and reduce to lowest terms.

6. Compare your answers from questions 4 and 5 and explain why any two slope triangles from any other line will behave the same as these two slope triangles.

7. Use the two numbers from your slope and either the word *increases* or *decreases* to complete the following statement to correctly describe the relationship demonstrated by the line: *For every* ___ *that x-increases, y* __________ *by* ____ *units.*

Proportional Relationships and Beyond

Until now, our real-world examples have focused on proportional relationships that cross through the origin at (0,0). For all of the relationships we have investigated, when the value of x was 0, the value of y was also 0. **Proportional relationships have the general form $y = mx$, where m is the slope.** In this section, we are going to briefly review writing equations in this form (we practiced this in Lesson 12, but we called the slope the *constant of proportionality*). Then we are going to progress onto lines that do not go through the origin, but instead have a non-zero value for y when $x = 0$. Those equations fall into the general form $y = mx + b$.

Proportional Relationships as y = mx

Proportional relationships have the general form $y = mx$, where m is the slope. Let's review some examples of situations that are appropriately represented in this format. Notice that in all these situations, when the independent variable is zero, the dependent variable is also zero:

- Edgar does not get a base salary at his sales job. Instead, he makes a 30% commission on the total amount of sound systems he sells. (If he sells nothing, he earns nothing.) *Pay = 30% × [sales]; P = 0.30s*
- Edgar's cat Shawnee needs to lose weight for her health. He puts her on a kitty treadmill to burn extra calories. For every minute she walks, Shawnee burns 1.5 calories. (If she does not go on the kitty treadmill, Shawnee does not burn any extra calories.) *Extra Calories Burned = 1.5 × [minutes]; c = 1.5m*
- Marsha is planting seeds from scratch that will grow two inches every week. (At zero weeks, the plants are zero inches tall.) *Height of Plant = 2 × [week #]; h = 2w*

In all of these examples, where the point (0,0) made sense, the rate of change can simply be plugged into the m value of $y = mx$.

Proportional relationships have the general form $y = mx$, where m is the slope.

Next we are going to look at how the equation $y = mx$ changes when the starting point of your y variable is *not* zero when $x = 0$.

Slope-Intercept Form: y = mx + b

It's official! Once we transition from the equations written as $y = mx$ to equations written as $y = mx + b$, we have made it to the big leagues! Why? Because lots of real-life mathematical relationships are not proportional. Remember, to be proportional the y term *must* equal zero when the x equals zero. Let's approach the equation $y = mx + b$ algebraically to determine what the value of y will be when $x = 0$. Sub in 0 for x and evaluate the equation for y:

$y = mx + b$; what happens when $x = 0$?
$y = m(0) + b$
$y = b$

Therefore, when $x = 0$, we see that $y = b$. This shows that all equations in the form $y = mx + b$ contain the point $(0,b)$. (This is a HUGE discovery!) We call b the *starting point* since it represents where the value of the mathematical relationship starts when $x = 0$. **Instead of crossing through the point (0,0), lines in the form $y = mx + b$ cross the y-axis at the number b.** The variable b is therefore also referred to as the *y-intercept* in the linear equation, $y = mx + b$. You should be familiar that both terms, *starting point* and *y-intercept*, refer to the value of b in the equation $y = mx + b$. Because b is the *y-intercept* and m is the *slope*, the equation $y = mx + b$ is called the *slope-intercept* form line.

In all linear equations written in slope-intercept form $y = mx + b$, b is the starting point. Since the line will cross the y-axis at the numerical value of b, b is also called the y-intercept. The variables m and b must always have numerical values and x and y must always be written as variables when expressing a relationship in slope-intercept form.

---

b: The Starting Point of y = mx + b

Now that you know that b is the starting point, let's discuss what that really means. The starting point is the y-value of a relationship at $x = 0$. Many relationships have a starting point that is something other than zero. Let's look at some examples of starting points:

- The base pay (starting salary) of a salesperson who earns additional commission according to their sales.
- The starting weight of a cat who is about to go on a diet.
- The existing mileage on a 1968 vintage Mustang's odometer before a new owner starts driving it 7,000 miles per year.
- The money that is already in Judy's bank account before she starts depositing $150 every month to save up for summer art classes.
- The starting height of a tree Ever plants that will grow 10 inches a year.

Practice 2

1. What is the difference between a proportional relationship and a linear relationship that isn't proportional?

2. What is the general equation of a relationship that is proportional? What point do *all* proportional relationships have in common on the coordinate plane?

3. What is the general equation of a relationship that is *not* proportional? What do all the different variables represent in this equation?

4. What point do *all* nonproportional math relationships have in common on the coordinate plane? (Hint: this point contains a variable.)

5. When a baby is born, an equation can be used to predict its weight over the first 50 months. Is this relationship proportional or nonproportional?

6. When mailing a package a courier service charges a flat "handling" fee and then an extra charge for every ounce a package weighs. Is this relationship proportional or nonproportional?

The Three Cases of Solving for Slope-Intercept Form

The goal in algebra is to transform given information into a rule, or mathematical equation, so that this equation can be used to identify more information or trends about the relationship. There are three different ways you will receive information, and in each case, you will be expected to model the relationship between two different variables as a mathematical equation. (*Rule* and *model* both refer to writing an equation in the form $y = mx + b$.) Here are the three different combinations of information you will be given:

> **Case 1:** You will be given the slope (rate) and the starting point (y-intercept).
> **Case 2:** You will be given the starting point and one coordinate pair, but not the rate.
> **Case 3:** You will be given two coordinate pairs, but not the rate or starting point.

Case 1: Using Starting Point and Slope to Write Equations

When you are given the *starting point* and the *rate of change* in a word problem, writing an equation to model the problem is as easy as plugging the *starting point* in for b and the *rate of change* in for m into $y = mx + b$. Make sure to leave x and y as variables since x is where you will input information in order to determine new values for y. Let's consider the growth of a baby bear.

> **Example:** A baby cub is born at nine inches long. If the cub is expected to grow two inches per week, write an equation to model the relationship between weeks and length of the bear.
> **Solution:** First, recognize that *time* is the independent variable and *length* is the dependent variable since *time* determines length. At zero weeks the bear is nine inches long so 9 is the starting point, or b. (In word problems you need to look for context clues to figure out the starting point.) The rate of change is the *two inches per week* (the word *per* indicates that it's the rate), so 2 is going to be the slope, m. If you want to change x and y to different variables so that they give more context to your equation, that's great! Just remember to define them. If w = number of weeks and l = length, then the equation $l = 2w + 9$ represents the relationship between time and length.

When given the slope and y-intercept that is not in the context of a word problem, writing an equation in the $y = mx + b$ form is even easier!

Example: What is the equation for a line that has a y-intercept of 4 and a slope of $\frac{2}{3}$?

Solution: Plug 4 in for b and $\frac{2}{3}$ in for m into the equation $y = mx +$ b: $y = \frac{2}{3}x + 4$.

Case 2: Using Starting Point and a Coordinate Pair to Write an Equation

Sometimes you'll be given just a starting point and enough information to make only one ordered pair of information. With proportional relationships, dividing the dependent variable by the independent variable gives you the rate of change, but this is not true when the starting point is not equal to zero! If you only have one ordered pair and the starting point, start with the general equation $y = mx + b$ and follow these four steps:

1. Plug your independent variable in for x and your dependent variable in for y
2. Plug your starting point in for b
3. Solve the equation for the slope, m
4. Rewrite the equation $y = mx + b$ with the correct values for m and b

Example: Forrest knows that it costs $5 to rent shoes at the bowling alley and then Mar Vista Lanes charges a fixed amount of money for every game he plays. On Saturday he rents shoes and bowls four games. His total bill when done is $23. Write an equation to model this information. How much did each game cost?

Solution: Since Forrest must pay $5 before he even starts bowling, that is his starting point, b. Then, since the price *depends* on the number of games, represent the other given information as the point (4,$23). We will sub b and the (x,y) coordinate pair into the slope-intercept equation and solve for the slope, m:

$$y = mx + b$$
$$23 = m(4) + 5$$
$$18 = 4m$$
$$\underline{\div 4 \quad \div 4}$$
$$4.5 = m$$

Now, since we want x and y to be variables in our slope–intercept equation, replace b with 5 and m with 4.5 to write the equation: $y = 4.5x + 5$. This equation shows that the starting point is $5 and the price per game is $4.50.

When given the y-intercept and a coordinate pair, writing an equation in the $y = mx + b$ form follows the same steps, but you might need to read the question carefully to understand that is the information you've been given:

Example: Write an equation that represents a line that crosses the y-axis at 8 and contains the point (2,4).
Solution: Start with the equation $y = mx + \text{b}$. Since the line crosses the y-axis at eight, plug 8 in for b. Then replace the x with 2 and the y with 4 so that you have enough information to solve for m:

$$y = mx + b$$
$$y = mx + 8$$
$$4 = m(2) + 8$$
$$\underline{-8 \qquad -8}$$
$$-4 = 2m \qquad \text{(divide both sides by 2)}$$
$$-2 = m$$

Then rewrite the equation $y = mx + b$ with the correct values for m and b: $y = -2x + 8$

Case 3: Writing Equations from Two Given Points

The last case is when you are not given the starting point *nor* the rate, but you are instead only given enough information to make two coordinate pairs. This is a most difficult case since you need to solve for the slope first, and then you must also solve for the y-intercept. These are the steps:

1. Represent the information as two coordinate pairs
2. Use the slope formula $m = \frac{y_2 - y_1}{x_2 - x_1}$ or a slope triangle on a graph to determine the slope.
3. Plug the slope in for m and the information from *either* ordered pair in for x and y in the equation $y = mx + b$, and then solve for b. (This is what you did in the previous method.)

Example: Kayla just started babysitting for a new family that pays her a flat fee for her travel time plus an hourly rate for her time spent watching the kids. She was too shy to get clarification of the rates her new emploers were paying her, but now that she's worked twice for them, she wants to try to figure out what her flat travel pay and hourly wage are. She made $46 for three hours of babysitting on Thursday and when she worked for $5\frac{1}{2}$ hours on Friday she received $76 from the family. Write an equation to model Kayla's pay scale and determine what her hourly rate and flat transportation pay are.

Solution: First, recognize that *hours* determine pay, so write your ordered pairs as (3hours,$46) and ($5\frac{1}{2}$ hours,$76). It is helpful to write in the units like "hr" and "$." Sub these coordinates into the slope formula and reduce your slope to lowest terms:

$$\textbf{Slope} = m = \frac{y_2 - y_1}{x_2 - x_1} = \frac{\$76 - \$46}{5.5 \text{ hours} - 3 \text{ hours}} = \frac{\$30}{2.5 \text{ hours}} = \frac{\$12}{1 \text{ hour}}$$

Since the slope is $\frac{\$12}{1 \text{ hour}}$, we have already determined that Kayla's hourly pay is $12. Next, we need to put 12 in for m, along with *either* of the coordinate pairs to solve for b. We will choose the point (3,46):

$$\begin{aligned} y &= mx + b \\ y &= 12x + b \\ 46 &= 12(3) + b \\ 46 &= 36 + b \\ -36 & \quad\; -36 \\ 10 &= b \end{aligned}$$

Now that you know that $b = 10$ and $m = 12$, plug those two values into the equation $y = mx + b$: $y = 12x + 10$. The word *flat* (*flat rate*, *flat fee*, *flat pay*) is usually a tip that this item is the starting point. Therefore, Kayla's flat transportation fee is $10, and her hourly pay is $12. It is important to mention that after you have solved for your slope, you can use either coordinate pair to help find the b in your equation. We happened to use the point (3,46), but now you should use the point ($5\frac{1}{2}$ hours,$76) along with a slope of $12, to see that this point would have resulted in the same value for b.

When this third case is presented without real-world context it will look like, "find the equation of the line that goes through the points (4,6) and (10,4)."

ERROR ALERT! Notice that the x value comes first in coordinate pairs, (x,y), but the y value comes first in the equation $y = mx + b$. This sets the stage for some serious mistake-making! When plugging a coordinate pair into the equation $y = mx + b$, it is way too easy to accidentally plug the x value in for y. Make sure you are careful with how you input coordinate pairs into $y = mx + b$!

Practice 3

1. Ruth bought a 1964 Corvair that had 120,000 miles on it. If her round-trip commute to work is 30 miles, write an equation to model the total mileage on the car after d days of driving to work.

2. Estes mailed a package to her friend in Lima, Peru, that was 12 pounds and cost $21. Then, the following month, she mailed another package to Peru that was eight pounds and cost $15. She knows the company charges a flat fee plus an additional cost per pound. Write an equation to model the fee structure that the shipping service charges to mail a package of p pounds to Peru.

3. Using your answer from question 2, what is the price per pound that the shipping service charges and what is their flat fee?

4. A line crosses the y-axis at negative two and has a slope of four. What is the equation to model this line?

5. After a birthday getaway weekend in El Segundo, Devon and Zaresh had to drive 360 miles to get back to their home in Vacaville. Write an equation to model the distance remaining if after three hours of driving they still have 201 miles left to go. How fast are they driving?

Answers

Practice 1

1. *Rise* represents the vertical *change in y-coordinates* and *run* represents the horizontal *change in x-coordinates*, so $\frac{\text{rise}}{\text{run}}$ is the same as $\frac{y_2 - y_1}{x_2 - x_1}$.
2. All slope triangles drawn on the same line are *similar triangles* and have proportional sides. When reduced to simplest form, the $\frac{\text{rise}}{\text{run}}$ slope produced by any two slope triangles from the same line will be equal.
3. The slopes between any collinear points will be equal.
4. 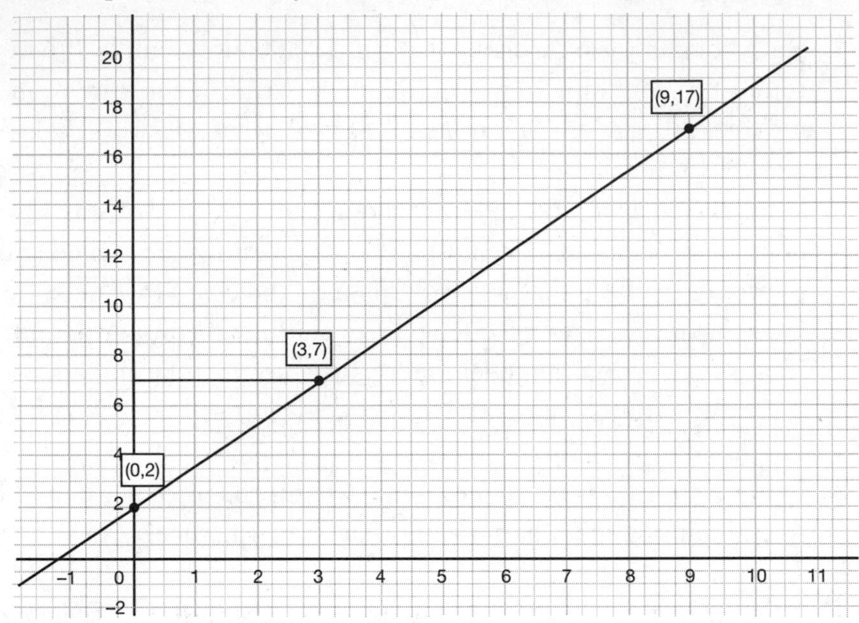

$\frac{\text{rise}}{\text{run}} = \frac{5}{3}$

5.

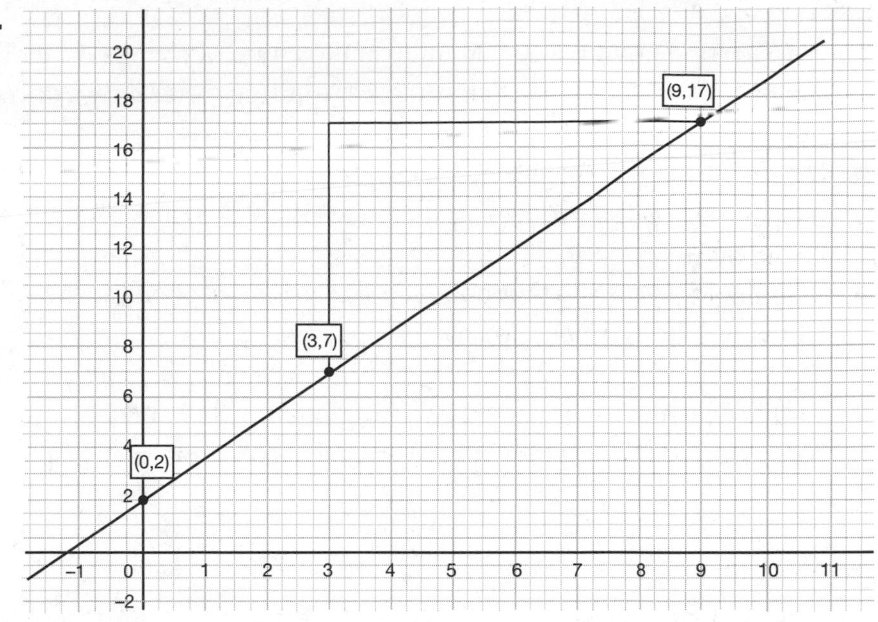

$\frac{\text{rise}}{\text{run}} = \frac{10}{6} = \frac{5}{3}$

6. The two different slope triangles produced equivalent slopes. This is true of any slope triangles drawn to the same line because slope triangles are always similar triangles that have proportional sides.

7. *For every <u>three</u> that x-increases, y <u>increases</u> by <u>five</u> units.*

Practice 2

1. A proportional relationship is a special kind of relationship that contains the point (0,0) and goes through the origin. Nonproportional relationships do not go through the origin.

2. $y = mx$ is the general equation for proportional relationships and they contain the point (0,0).

3. $y = mx + b$ is the general equation for nonproportional linear relationships. m represents the slope, b represents the starting point or y-intercept, and x and y represent (x,y) coordinate pairs that sit on the line.

4. *All* nonproportional relationships contain the point $(0,b)$, which is the starting point or y-intercept.

5. Since the baby's starting weight is not zero at day zero, this relationship is not proportional.

6. Since there is a flat handling fee to begin with, this relationship is not proportional.

Practice 3

1. The starting point is 120,000 miles and the rate of change is 30 miles per day. $y = 30d + 120{,}000$ represents the mileage after d days of commuting to work.
2. (12,\$21) and (8,\$15). The rate of change is $\frac{3}{2}$. And when this is subbed back into $y = mx + b$ along with the point (12,\$21) we find that b is five. So $y = \frac{3}{2}x + 5$.
3. The flat fee is the starting point, 5. The price per pound is the slope, which is $\frac{3}{2}$ or \$1.50 per pound.
4. $y = 4x + -2$ or $y = 4x - 2$
5. In this equation we are given a starting point of 360 and a coordinate pair of (3,201). Plug these into $y = mx + b$ to solve for m:

$$201 = 3m + 360$$
$$\underline{-360 \qquad -360}$$
$$-159 = 3m$$
$$-53 = m$$
$$y = -53m + 360$$

Notice that in this equation the slope is –53. This represents that their distance from home is *decreasing* at a rate of 53 miles per hour—they must be traveling at 53 miles per hour!

23

Solving Multi-Step Linear Equations

STANDARD PREVIEW

In this lesson we will cover **Standards 8.EE.C.7.A** and **8.EE.C.7.B.** You will learn how to identify when linear equations have one, many, or no solutions. You will also learn how to solve complex multi-step linear equations.

Equations with One, Infinite, or No Solutions

One day your friend walks into class and tells you he is going to ask you three riddles, and he wants you to determine which riddle has just one solution, which riddle has infinite solutions, and which riddle has no solutions. Here are his three (bad) riddles:

> **Riddle A:** "On this day, I will be exactly 14 years old. What day am I?"
> **Riddle B:** "On this day, Earth will be on an orbit around the sun. What day am I?"
> **Riddle C:** "On this day, my dog Pepe will prepare and serve our algebra class Chicken Cordon Bleu. What day am I?"

It's pretty easy for you to draw the following conclusions:

> **Riddle A** has *exactly 1 solution*: your friend's 14th birthday.
> **Riddle B** has *infinite solutions*: the Earth is on an orbit around the sun *every* day.
> **Riddle C** is kind of creepy to even think about, but it obviously has *no solution* since no matter how smart and talented Pepe is, a dog is not going to prepare and serve a class of teens Chicken Cordon Bleu.

Up until now our focus has been on linear equations that have had one unique solution. Now you are old enough to know that there are also single-variable equations whose solutions fall into the other two categories: infinite solutions and no solutions. Here are examples of single-variable equations that have one, infinite, and no solutions:

> **Equation A:** $x + 7 = 10$
> **Equation B:** $x + 7 = x + 6 + 1$
> **Equation C:** $x = x + 7{,}000$

Equation A, $x + 7 = 10$, has just one value of x that makes the equation true: $x = 3$.

Equation B, $x + 7 = x + 6 + 1$, has infinite solutions since no matter what x is on the left side of the equation, when we add it to 7 we will end up getting the same thing as when we add that x to six and then to one on the right side of the equation. *When an equation has infinite solutions, it means no matter what value is plugged in for the unknown variable, the final equation will be a true numerical statement.*

On the other hand, Equation C, $x = x + 7{,}000$, will never be true for any value of x. There is no number x that, when increased by 7,000, is still the same as its original value. For example, when we plug in $x = 10$, we'll get $10 = 7{,}010$, which is a false statement. Just thinking about this might make you kind of dizzy because it just doesn't make any sense! *When an equation has no solutions, it means no matter what value is plugged in for the unknown variable, the final equation will be a false numerical statement.*

The Three Simplified Forms of Single-Variable Equations

There is a trick to knowing if a single-variable equation will have one, infinite, or no solutions. No matter how long and confusing an equation may be, it will be possible to simplify it to one of three basic forms. These unique forms tell us which category an equation falls into. The three basic forms are:

- $x = a$, where a is a constant and x is a variable
- $a = a$, where a is a constant that equals itself
- $a = b$, where a and b are two different constants

x = a: Equations with One Solution

We have already had practice solving equations for a single value that makes the equation true. We kept the equation balanced by doing the same operation to *both* sides of the equation. We also used opposite operations to get the variable alone on one side of the equation, while a numerical value was on the other side. The equation $x + 7 = 10$ is a simple example of this type of equation. After subtracting seven from both sides we are left with the equation $x = 3$. This equation is now in the $x = a$ form, where a is a constant and x is the variable. This final form always indicates that there is just one solution to an equation.

When Your Variable Goes Missing

As we enter new territory, where there can be *infinite* or *no* solutions, you will notice that sometimes the variable term will disappear entirely. When this happens, there's no need to call a detective—it is just an indication that this equation either has *infinite* or *no* solutions. Since *infinite solutions* is very different from *no solutions*, let's learn how to distinguish between these two cases!

a = a: Equations with Infinite Solutions

Consider the equation $x + 7 = x + 6 + 1$. When an equation has a variable on both sides, we need to collect all the variable terms on one side of the equation and move all the constants over to the opposite side of the equation. We still do this by using opposite operations. Watch what happens as we subtract the x term from the right-hand side to bring it over the left-hand side:

$$\begin{array}{rl} x + 7 &= x + 6 + 1 \\ \underline{-x \quad\;\;} & \underline{\;-x \quad\quad\;\;} \\ 7 &= 6 + 1 \\ 7 &= 7 \end{array}$$

The x terms in the equation above have completely disappeared and we are left with an equation, $7 = 7$. x was canceled out by $-x$. Let's think about what this numerical equation is now stating: it states that *seven equals seven*. When is that statement true? **All** the time. When we end up with a true statement in the form $a = a$, where a is a constant that equals itself, this means that the equation has *infinite solutions*. No matter what value is subbed in for x, the resulting equation will be a true statement.

In general, equations will have infinite solutions if the left-hand side of the equation is an expression that is equivalent to the right side of the equation. For example, since $5(2x + 3)$ is an equivalent expression to $10x + 15$ (the 5 has been distributed), this equation will have infinite solutions: $5(2x + 3) = 10x + 15$.

a = b: Equations with No Solutions

Consider the equation $x = x + 7{,}000$. Since this equation has a variable on both sides, let's collect the x terms on one side and keep the numerical terms on the other side. We do this by using opposite operations and again, our variable is going to cancel out:

$$\begin{array}{rl} x &= x + 7{,}000 \\ \underline{-x} & \underline{\;-x \quad\quad\quad} \\ 0 &= 7{,}000 \end{array}$$

When the x terms in the equation cancel each other out, we are left with the numerical equation, $0 = 7{,}000$. Let's think about what this equation is stating: it states that *zero equals seven thousand*. When is that statement true? **Never!** This is a wildly false statement! When we end up with a false statement in the form $a = b$, where a and b are two different constants, this false claim signifies that the original equation has *zero solutions*. No matter what value is subbed in for x, the resulting equation will be a false statement.

In general, equations will have no solutions when it is impossible for the expression on the left to ever equal the expression on the right. Look for equations that have equivalent variable terms on both sides, but different constant terms. For example, $14x + 2 = 14x + 10$ can never have a solution because the $14x$ terms will always be the same value but the 2 and the 10 will always make the two sums different.

Practice 1

For questions 1 through 3, determine whether the following statements are sometimes, always, or never true:

1. A single-variable equation has one unique solution.

2. When the variable cancels out in an equation, there will be no solutions to the original equation.

3. A final numerical statement in the form $a = b$ where both a and b are constants, is an indication that the original equation does not have any solutions.

4. How is it possible that an equation can have infinite solutions? Write an example of such an equation that has $4(x + 2)$ on the left-hand side.

5. How is it possible that an equation can have no solutions? Write an example of such an equation that has $2x + 5$ on the left-hand side.

6. Does the equation $2x + 60 = 2x + 600$ have one, infinite, or no solutions?

7. Does the equation $2(3x + 12) = 3(2x + 8)$ have one, infinite, or no solutions? (Hint: It will help if you simplify both sides of the equation first.)

Solving Multi-Step Linear Equations

The first types of equations we practiced solving in Lesson 9 were one-step equations in the form $7.9 = -4.1 + x$ or $\frac{8}{5}x = 40$. To solve these equations you used just *one* opposite operation, like addition or division, to get x alone. In Lesson 17, we reversed the order of PEMDAS to solve two-step equations that look like $-8x - 10 = -22$. To solve these equations we needed to use addition or subtraction *and* division to get x alone. In this section, we are going to take things up another notch by working with equations with multiple x-terms. We'll also work with problems that have parentheses that require the distributive property.

Equations with Multiple x-Terms

A *single-variable equation* doesn't necessarily have just one term with a variable—a *single-variable equation* means that an equation doesn't have *two different variables*. $y = 4x + 5$ is not a single-variable equation, but $4x + 10 = 2x - 20$ is a single-variable equation. The equation $4x + 10 = 2x - 20$ brings us to our next task at hand: solving equations that have more than one x-term.

When an equation has one or multiple variable terms on both sides of the equal sign, there will just be one or two additional steps to solving your equation. (We still suggest first rewriting equations with subtraction as addition by using *keep-switch-switch*):

Step 1: Combine like terms on the same side of the equation by using the operations associated with them.
Step 2: Move the like terms to one side of the equation by using opposite operations. (It doesn't matter if you move the variable terms to the left side or the right side. Some students like to always move the variable to one specific side, while other students prefer to move the variables in a way that will keep them positive.)

Then we will solve the equation like a two-step equation: we will move the constant terms to the *other* side of the equation and then get the variable term alone by dividing both sides by its coefficient.

ERROR ALERT! Do *not* use opposite operations to combine like terms on the *same side* of the equation. Use opposite operations only to move the variable term *to the other side of the equation*. Students get nutty and start using opposite operations on the same side of an equation once they start seeing more than one variable. Don't do that! Simplify the following equation by *adding* the $10x$ and $2x$, which are both on the same side of the equation. Do not use opposite operations until you are moving variable terms *to the other side* of an equation!

$\underline{10x} + 9 \underline{+ 2x} = \underline{4x} + 5$ ($10x$ and $2x$ must be combined)
$\underline{12x} + 9 = 4x + 5$ (We correctly added $10x$ and $2x$)
$8x + 9 = 4x + 5$ (We incorrectly used opposite operation $-2x$ to combine it with the $10x$)

Now let's work through an example of a multi-step equation where you'll have to combine like terms on one side of the equation and then use opposite operations to move them to the other side of the equation:

Example: What value of x makes the equation true: $4x + 10 = 5x - 20 - 3x$?
Solution:

$4x + 10 = 5x - 20 - 3x$	(Apply keep-switch-switch)
$4x + 10 = \underline{5x} + -20 + \underline{-3x}$	(Underline like terms on right side)
$4x + 10 = 2x + -20$	(Combine $5x + -3x$ on right side)
$\underline{-2x \qquad -2x}$	(Subtract $2x$ from both sides)
$2x + 10 = -20$	
$\underline{\qquad -10 \quad -10}$	(Move 10 to opposite side of $2x$)
$\dfrac{2x}{2} = \dfrac{-30}{2}$	(Divide both sides by 2 to get x alone)
$x = -15$	(Arrive at final answer)

So, we have arrived at –15 being the value for x that makes the equation true. Now we will check our answer by plugging –15 in for *all* the values of x:

$4x + 10 = 5x - 20 - 3x;\ x = -15?$	
$4(-15) + 10 = 5(-15) - 20 - 3(-15)$	(Plug –15 in for all x values)
$-60 + 10 = -75 - 20 - (-45)$	(Multiplication)
$-50 = -75 + (-20) + 45$	(Keep-switch-switch and begin simplifying)
$-50 = -95 + 45$	(Addition)
$-50 = -50$ ✔	(Arrive at true statement. –15 must be correct!)

Equations with Parentheses

In an earlier lesson we learned that we could solve equations like $4(c - 3) = 40$ by treating the $(c - 3)$ as a single term so that we could just divide both sides by 4:

$$4(c - 3) = 40$$
$$4(\blacksquare) = 40$$
$$\div 4 \qquad \div 4$$
$$(\blacksquare) = 10$$

So $c - 3 = 10$, and $c = 13$

This technique should not to be used to solve more complicated equations where there is more than one term on either side of the equation. In the case of the following equation, it is necessary to distribute the 2 before proceeding with the question. We left the steps blank for you to fill in—what is being done each step? Check your answers below the example.

Example: One Solution

$2(x - 6) = 4x + 20$	
$2(x + -6) = 4x + 20$	1. (____________________)
$2(x) + 2(-6) = 4x + 20$	2. (____________________)
$2x + (-12) = 4x + 20$	3. (____________________)
$-2x \qquad -2x$	4. (____________________)
$-12 = 2x + 20$	5. (____________________)
$-20 \qquad -20$	6. (____________________)
$-32 = 2x$	7. (____________________)

$\div 2 \quad \div 2$ 8. (____________________)

$-16 = x$ 9. (____________________)

So the one value of x that will make this equation truc is -16. Here are the steps we did to arrive at this answer: (1) apply *keep-switch-switch*; (2) distribute; (3) multiply; (4) move x-terms to the right using opposite operations (we moved them to the right so that x would stay positive, but you'd still get the correct answer if you moved the x-terms to the left); (5) combine like terms (subtact); (6) move constants to side opposite x with opposite operations; (7) subtract; (8) divide both sides by 2; (9) arrive at final answer!

Here is a review of the rules we covered in the first half of this lesson. Read them carefully before moving on to the last examples in this lesson.

All single-variable equations can be simplified to one of three basic forms. Each form is a key that reveals how many solutions exist for that equation:

Case 1: ***x = a, where a is a constant, determines there is one solution.***

- **This solution will look like $x = \frac{1}{2}$ or $h = 5$.**

Case 2: ***a = a, where a is a constant, determines there are infinite solutions.***

- **This solution will look like a true numerical equation like $-5 = -5$ or $0 = 0$**

Case 3: ***a = b, where a and b are two different constants, determines there is no solution.***

- **This solution will look like a false numerical equation like $5 = -5$ or $0 = 4$**

Example: No Solutions

$2(6x - 8) = 10 - 4(-3x - 5)$

$2(6x + -8) = 10 + (-4)(-3x + -5)$ (*Keep-switch-switch*)

$2(6x) + 2(-8) = 10 + (-4)(-3x) + (-4)(-5)$ (Distribute)

$12x + (-16) = 10 + 12x + 20$ (Multiply)

$-12x \qquad -12x$ (Bring x-terms onto same side)

$-16 = 32$ (Combine like terms)

Since $-16 = 32$ is a false equation that is never true, there is no solution to the equation $2(6x - 8) = 10 - 4(-3x - 5)$.

Example: Infinite Solutions

$10x + 4 - 2x = 2(4x - 5) + 14$

$10x + 4 + (-2x) = 2(4x + -5) + 14$ (*Keep-switch-switch*)

$10x + 4 + (-2x) = 2(4x) + 2(\ 5) + 14$ (Distribute)

$\underline{10x} + 4 + (-2x) = 8x + \underline{(-10)} + \underline{14}$ (Multiply)

$8x + 4 = 8x + 4$ (Combine like terms on both sides)

$\underline{-8x \qquad -8x}$ (Bring x-terms onto same side)

$4 = 4$ (Combine like terms)

Since $4 = 4$ is a true equation that is always true, there are infinite solutions to the equation $10x + 4 - 2x = 2(4x - 5) + 14$.

Practice 2

1. What additional steps will you have to do when solving multi-step linear equations that have more than one variable, such as the equation $6x + 36 = 8x + 20 - 6x$. Explain the necessary extra steps.

2. Solve the equation from question 1.

3. When solving equations with variable terms on both sides is it necessary to always move the variable terms to the left-hand side? Explain why this is or may not be the best method.

4. What is a common mistake that students make when there is more than one variable term on the same side of the equation? Use the equation $4x - 7 = 5x + 4 + 3x$ to illustrate the incorrect steps that students tend to make. Then solve the equation correctly.

5. Solve the equation $\frac{1}{2}(8x + 10) = 2 + 4x + 3$

6. Solve the equation $\frac{2}{3}(3x + 12) = 4(x + 3) - 2x$

Answers

Practice 1

1. A single-variable equation <u>*sometimes*</u> has one unique solution. It can also have no solution or infinite solutions.
2. This statement is sometimes true. It depends on if the resulting numerical statement is true or false.
3. A final numerical statement in the form $a = b$ where both a and b are constants, is <u>*always*</u> an indication that the original equation does not have any solutions.
4. Equations have infinite solutions when the left-hand side of the equation is an expression that is equal to the right side of the equation. $4(x + 2) = 4x + 8$ will have infinite solutions because no matter what values you put in for x, both sides of the equation will be equal.
5. An equation will have no solutions when it is impossible for the expression on the left to ever equal the expression on the right. For example, adding 1,000 to $2x$ can never be the same as adding five to $2x$, so the expressions $2x + 5 = 2x + 1{,}000$ will have no solutions.
6. $2x + 60 = 2x + 600$ has no solutions.
7. $2(3x + 12) = 3(2x + 8)$ has infinite solutions since both sides are equivalent to $6x + 24$.

Practice 2

1. When starting with $6x + 36 = 8x + 20 - 6x$:

   Step 1: Combine like terms on the right side of the equation: $6x + 36 = 2x + 20$

   Step 2: Move variable terms over to the same side of the equation using opposite operations:

   $$\begin{array}{l} 6x + 36 = 2x + 20 \\ \underline{-2x \qquad\quad -2x} \\ 4x + 36 = 20 \end{array}$$

2. $6x + 36 = \underline{8x} + 20 \underline{- 6x}$

   $6x + 36 = \underline{8x} + 20 \underline{+ (-6x)}$

   $6x + 36 = 2x + 20$

   $\underline{-2x \qquad -2x}$

   $4x + 36 = 20$

   $4x \qquad = -16$

   $x \qquad = -4$

3. It is not necessary to move the variable terms to either one side or the other. Sometimes it is easiest to move the variable terms in a manner that keeps them from becoming negative, but this is a personal preference and is not required. You will arrive at the same answer, regardless of what side you move the variable terms to, as long as you perform opposite operations consistently when moving terms.

4. With an equation like $4x - 7 = 5x + 4 + 3x$ students tend to use opposite operations when combining terms on the same side of the equations. Instead of combing $5x$ and $3x$ to get $8x$, students might make the error of doing $5x - 3x = 2x$, even though $5x$ and $3x$ are on the same side of the equation. Only use opposite signs when moving terms to the other side of the equation.

$$4x - 7 = 5x + 4 + 3x$$
$$4x - 7 = 8x + 4$$
$$\underline{4x - 11 = 8x}$$
$$-11 = 4x$$
$$-\frac{11}{4} = x$$

5. $\frac{1}{2}(8x + 10) = 2 + 4x + 3$

$$4x + 5 = 2 + 4x \underline{+ 3}$$
$$4x + 5 = 4x \underline{+ 5}$$
$$\underline{-4x \qquad -4x}$$
$$5 = 5$$

This equation has infinite solutions since it simplified to a true numerical statement.

6. $\frac{2}{3}(3x + 12) = 4(x + 3) - 2x$

$$\frac{2}{3}(3x) + \frac{2}{3}(12) = 4x + 12 - 2x$$
$$2x + 8 = \underline{4x} + 12 \underline{+ (-2x)}$$
$$2x + 8 = 2x + 12$$
$$\underline{-2x \qquad -2x}$$
$$8 = 12$$

This equation has no solutions since it simplified to a false numerical statement.

24 Understanding Systems of Equations

STANDARD PREVIEW

In this lesson we will cover Standards 8.EE.C.8.A and 8.EE.C.8.B. What is significant about the point where two linear equations on the same coordinate plane intersect? What does it mean when two lines on the same coordinate plane are parallel? You will learn the answers to these questions along with how to solve for the one coordinate pair that solves two linear equations at the same time.

Understanding Systems of Equations

A system of equations is when two linear equations are considered at the same time. Solving a system of equations is similar to meeting two people's different requirements at once. We meet multiple people's requirements in life all the time. At least Alex does . . . She comes home from college to Kokomo, Indiana, for the summer and tells her parents she wants to backpack around Europe for a few months. Her mom says, "I don't really care *what* you do this summer, I just want you to live at home." Her dad chips in, "I don't really care *where* you are this summer, I just want you to have a job and earn some money for college." In order to satisfy *both* her parents, Alex has to get a job in Kokomo this summer and save her European backpacking fantasy for another summer!

It sounds so complicated and scary! But really, *solving a system of equations* just means finding the (x,y) coordinate pair that works in two different equations at the same time. For example, the point (3,2), a solution to the equation $x + y = 5$, also solves the equation $x - y = 1$. Therefore, we would say that *(3,2) is a solution to the system of equations* $x + y = 5$ *and* $x - y = 1$. See—that wasn't too bad! Before we discuss *how* to find a solution to a system of equations let discuss what we're really looking at when we see a line in a coordinate plane.

Seeing Lines as Infinite Solutions

We have learned how to graph linear equations by plotting ordered (x,y) pairs on a coordinate plane and connecting the points. We know those ordered pairs we plotted represent solutions to the equation, but what do the lines themselves represent? *Every line in the coordinate plane is a collection of infinite coordinate pairs that make that equation true.* Let's consider the equation $x + y = 5$. Several obvious solutions to this equation are listed in the following table:

x	y
0	5
1	4
2	3
3	2

When we plot these points on the graph we have four distinct solutions:

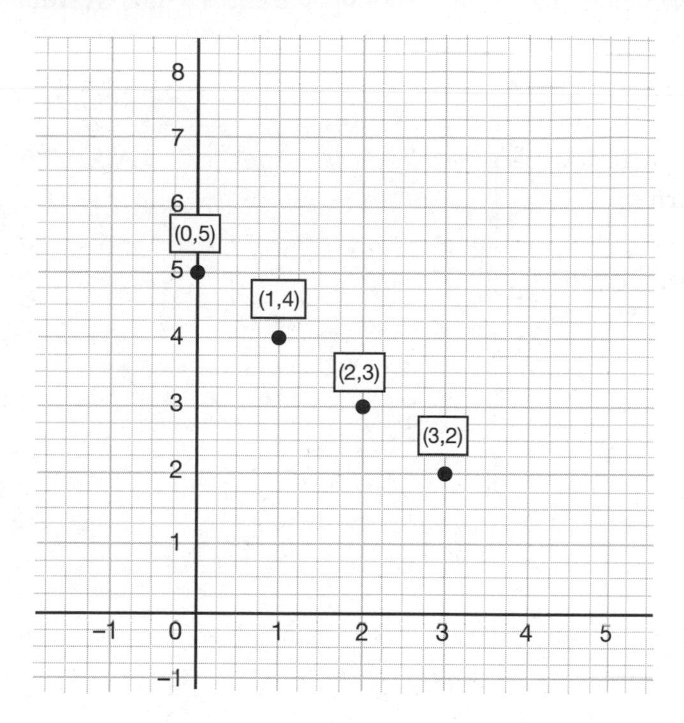

But once our graph goes from a collection of points to a line connecting those points, what are we really indicating? We are actually filling in all the coordinate pairs that work in the equation. These points didn't make it into our original table, but they will exist on the line once those original points are connected. Let's take a look at a few examples of the types of coordinate pairs that are on this line but weren't in our original table:

x	y
0.1	4.9
0.2	4.8
0.61	4.39
$\frac{2}{3}$	$4\frac{1}{3}$

"Eww, those are some ugly points," you might be saying to yourself, but the equation $x + y = 5$ doesn't think so! Every equation has lots of decimal and fraction points that often get left out of tables but are still valid solutions to the equation. A line on a coordinate plane *includes* all those previously overlooked solutions. How many solutions do you think are shown on our graph now that we have connected the original unconnected four coordinate pairs?

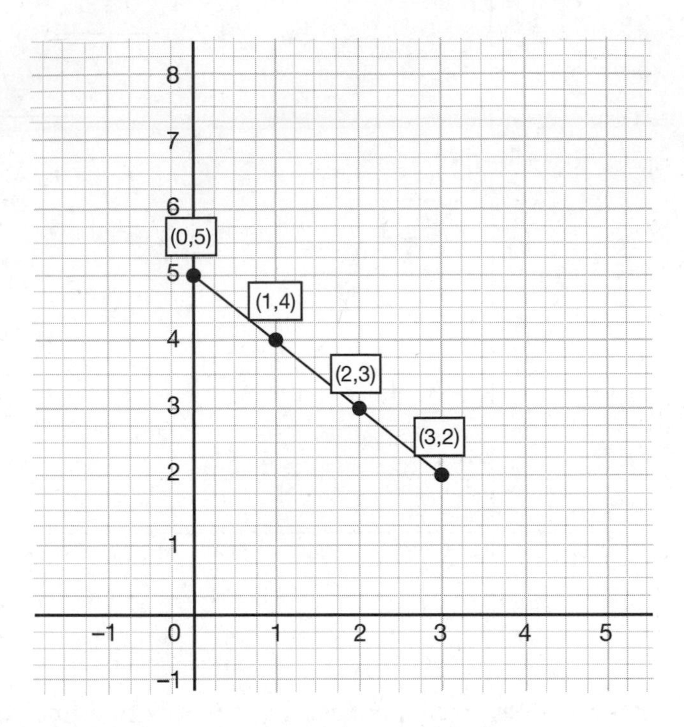

Hopefully, you see that the line segment shown illustrates an *infinite* number of coordinate pair solutions! Every line, or even a segment of a line, has infinite solutions; there are countless decimal values between any two numbers. It's okay if this "infinite solutions thing" is a little mind-blowing. We know. Give it some time.

Understanding the Point of Intersection

Now that you understand that a line is a collection of infinite points, let's look at what it means to have two lines graphed on the same coordinate plane. We'll put our original two equations, $x + y = 5$ and $x - y = 1$, on the same graph. Let's start by making tables of ordered pairs for each equation:

$x + y = 5$

x	y
1	0
2	1
3	2
4	3

$x - y = 1$

x	y
0	5
1	4
2	3
3	4

Now we'll put them both on the same coordinate plane and connect each series of points to make two lines:

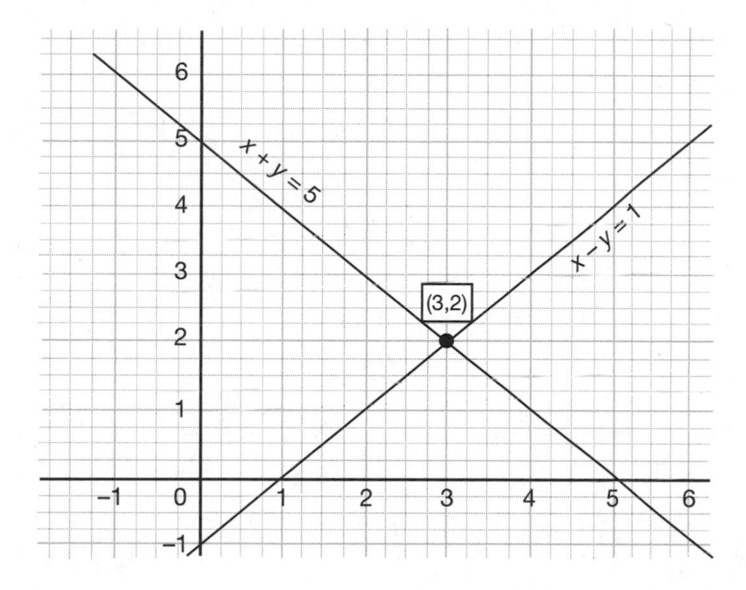

Each line is showing a collection of infinite solutions to the linear equation that represents the line. Therefore, what can you determine about the point where the two lines cross? What is that coordinate pair? Can you find that same coordinate pair in both tables above? The point where two lines intersect marks the (x,y) coordinate pair that is a solution to both linear equations. It is called *the point of intersection* and is the solution to the system of equations. Notice that the lines cross only at one point, (3,2), and that this point is in both of the tables.

Congratulations! You have just learned how to solve a system of linear equations through graphing. We thought it would be better to just spring this on you, since we realize it sounds much scarier than it actually is! Here are the official steps you can follow to solve other systems of equations by graphing:

> **Step 1:** Make tables of values for both linear equations. In each table, plug in at least three values for x and then solve each equation for three corresponding y-values.
> **Step 2:** Plot the coordinate pairs from one table on a coordinate plane and connect them to make a line. Repeat for the second table.
> **Step 3:** Identify the coordinates of the ordered pair that marks the spot of intersection of the two lines.
> **Step 4:** Check your solution. Make sure that this coordinate pair is in both tables or plug the point (x,y) into both linear equations to verify that it satisfies both equations.

Practice 1

1. What is a system of equations?

2. Line segment $\overline{AB}$ connects point A (1,5) to point B (4,5). How many coordinate pairs exist on this line segment?

3. How can a graph be used to find a solution to a system of equations?

4. What does it mean if a point (1, 9) is a solution to a system of equations?

5. Use the following tables and coordinate plane to solve the system of equations by graphing:

$y = -x + 9$

x	y

$y = x + 1$

x	y

6. Use the following tables and coordinate plane to solve the system of equations by graphing. (Hint: Since the slopes are fractions, choose numbers for x carefully that will help cancel out the denominators—we helped you with one suggestion in each table.)

$y = \frac{1}{2}x + 4$

x	y
8	

$y = -\frac{1}{3}x + 9$

x	y
6	

7. Check your answer to question 6. Plug the coordinate pair you identified as the point of intersection into the original equations to see if it is a solution for both equations.

Solving Systems of Equations Algebraically

Sometimes graphing two equations isn't the most efficient way to solve a system of equations. What if your equation has extremely large numbers that won't easily fit on a graph? What if two lines intersect at a fractional coordinate pair like $(\frac{1}{3}, 2\frac{1}{8})$? That wouldn't be too easy to read on a graph! Other times it might just be a hassle to make tables and a coordinate graph in order to find the solution. We know you might be hesitant to believe us, but sometimes it's just easier to solve an equation using some algebra and opposite operations!

There are three different ways to solve systems of equations using algebra instead of graph paper. We are going to let you know when each method is most efficient (what to look for in the equations) and then give you an example of how to use this method. It's important to note that although you are accustomed to seeing equations presented in the $y = mx + b$ format, linear equations come in all different flavors and can be presented in a variety of ways. For example, the format $Ax + By = C$, where A, B, and C are all non-zero numbers, is called *standard form*. (You will see in the next lesson that this form is quite helpful to represent certain real-world contexts.)

The Common Thread of Algebraic Solutions

All equations with two variables have *infinite* pairs of solutions. A system of equations is two lines that *both* have infinite pairs of solutions. That's a lot of solutions to sort through! So what is the underlying technique that the three algebraic methods depend on to solve a system of equations? As you will see, each of these approaches relies on *combining the two equations in a way that one of the variables is canceled out*. That way, a single-variable equation remains and can be solved. Like we learned in the previous lesson, all single-variable equations have either one, none, or infinite solutions so we will therefore simplify the new single-variable equation to see if it falls into the $x = a$, $a = a$, or $a = b$ case.

The Equal Values Method

You know how to solve a system of equations by graphing, but what if you were given the equations $y = 2x + 100$ and $y = 8x - 140$? Those are too big to graph! *The Equal Values Method* is the quickest way to solve two large–number equations that are both in the form $y = mx + b$. Since $y =$ *Equation 1* and $y =$ *Equation 2* we can set *Equation 1* equal to *Equation 2*. Then we will have gotten rid of the y terms and we can solve a single variable equation:

Example: Find the solution that satisfies both $y = 2x + 100$ and $y = 8x - 140$

Solution:

Step 1: Set the equations equal to each other and solve for x:

$$2x + 100 = 8x - 140$$
$$\underline{-2x \qquad\quad -2x}$$
$$100 = 6x - 140$$
$$\underline{+140 \qquad\quad +140}$$
$$240 = 6x$$
$$\underline{\div 6 \quad \div 6}$$
$$40 = x$$

Step 2: Now that we have $x = 40$, we want to know what y equals when $x = 40$. Plug 40 in for x into *either* of the original equations:

$y = 2x + 100$; $x = 40$

$y = 2(40) + 100$

$y = 80 + 100$

$y = 180$, so (40,180) is the solution.

Step 3: Now that we have (40,180) as our solution, plug this point into whichever equation you *didn't use* in step 2 to check the answer:

$y = 8x - 140$; for (40,180)
$180 = 8(40) - 140$
$180 = 320 - 140$
$180 = 180$ ✔

Step 4: The point (40,180) satisfies both equations and is the solution to our system.

When using any of the algebraic methods to solve systems of equations, you will follow these four general steps:

1) **Combine the two equations in a way that eliminates one of the variables so you can solve for either *x* or *y*.**
2) **Plug that *x* or *y* value into one of the original equations to solve for the other variable.**
3) **Check your coordinate pair solution by plugging it into the *other* equation that wasn't used in step 2 to make sure it satisfies the second equation.**
4) **Represent your solution as a coordinate pair with the *x*-value first.**

The Substitution Method

The Equal Values Method is only easiest when both equations are in the same exact format and both equal the same value. When you are handling equations that are written in two different formats, consider *The Substitution Method.* This approach is particularly useful when *one of the variables is by itself.* That variable's equivalent expression can be substituted in for that same variable in the other equation in order to create a single-variable equation:

Example: What coordinate pair is the solution to $y = 2x - 4$ and $6x - 4y = -2$?

Solution:

Step 1: Since $y = 2x - 4$ in the first equation, replace the y in the second equation with $2x - 4$ in order to create a single-variable equation:

Using $6x - 4\underline{y} = -2$, plug $y = \underline{\mathbf{2x - 4}}$ in for y:

$6x - 4(\underline{2x - 4}) = -2$

$6x + (-4)(2x + -4) = -2$ (Rewrite subtraction as addition when distributing!)

$6x + (-4)(2x) + (-4)(-4) = -2$

$6x + \quad (-8x) + \quad 16 \quad = -2$

$-2x + \quad 16 \quad = -2$

$\underline{-16 \quad -16}$

$-2x \quad = -18$

$\underline{\div(-2) \quad \div(-2)}$

$x = 9$

Step 2: So, if $x = 9$, what is the y-coordinate that corresponds with that x-coordinate? To find y, we substitute 9 in for x into either of the original equations.

$y = 2x - 4$

$y = 2(9) - 4$

$y = 18 - 4$

$y = 14$; (9,14) is the solution

Step 3: Now that we have (9,14) as our solution, plug this point into whichever equation you *didn't use* in step 2 to check the answer:

$6x - 4y = -2$; for (9,14)

$6(9) - 4(14) = -2$

$54 - 56 = -2$

$-2 = -2$ ✔

Step 4: The point (9,14) satisfies both equations and is the solution to our system.

The Elimination Method

Commonly called *Linear Combination* or *The Addition Method*, *The Elimination Method* is a useful approach when both equations are presented in standard form, $Ax + By = C$. As its names suggest, we will be *adding* the two equations together in order to *eliminate* one of the variables. The goal is to use algebra to manipulate the two equations so that either the x-terms or the y-terms have opposite coefficients and will cancel each other out when added (like $8x$ and $-8x$).

Example: Find the solution to the linear system $3x + 2y = 46$ and $2x + 6y = 68$.

Solution:

Step 1: First, stack the two equations to clearly compare them:

$3x + \underline{2y} = 46$

$2x + \underline{6y} = 68$

Then, determine if either of the x-terms or y-terms from one equation can be multiplied to make it the opposite value of its like term in the other equation. We underlined the y-terms because $2y \times (-3) = -6y$. (If we can change the y-term in the first equation to $-6y$, then it will cancel out the $6y$ in the second equation when we add them.)

Step 2: Multiply *both sides* of one entire equation by the factor identified in Step 1 to create an equivalent equation:

$-3(3x + 2y) = (46)(-3)$

$-9x + -6y = -138$

This equation has a y-term with an opposite coefficient to the y-term in the other equation, so when we add them together, the y-terms will cancel.

Step 3: Stack the new equation with the other original equation and add them together. Then solve for the remaining variable:

$$\begin{array}{rcl} -9x + -6y & = & -138 \\ +\underline{(2x + \ 6y} & \underline{=} & \underline{\ 68)} \\ -7x + 0 & = & -70 \\ \underline{\div(-7)} & & \underline{\div(-7)} \\ x & = & 10 \end{array}$$

Step 4: Now we need to find the y-coordinate that corresponds with $x = 10$. To find y, we plug 10 in for the x using either of the original equations.

$$\begin{array}{rcl} 3x + 2y = 46; & x = 10 & \\ 3(10) + 2y & = & 46 \\ 30 + 2y & = & 46 \\ \underline{-30} & & \underline{-30} \\ 2y & = & 16 \\ y & = & 8 \end{array}$$

Step 5: Now that we have (10,8) as our solution, plug this point into whichever equation we *didn't use* in step 2 to check the answer:

$$2x + 6y = 68; \text{ for } (10,8)$$
$$2(10) + 6(8) = 68$$
$$20 + 48 = 68 \checkmark$$

Step 6: The point (10,8) satisfies both equations and is the solution to our system.

Since *The Elimination Method* requires additional set-up in order to get one of the terms ready to cancel out the other, notice that it has an additional two steps. Sometimes there will be even one more step if *both* of the equations need to be multiplied in order to create opposite coefficients to cancel out. Consider the two following equations:

$$3x + 2y = 10$$
$$4x + 5y = 18$$

Similar to how we find *common denominators* when adding fractions, both equations will need to be written as equivalent equations so that the x-terms or y-terms will cancel out. Let's set these equations up to cancel out the x terms by making the x-coefficient 12 in one equation and –12 in the other equation:

$$4 \times (3x + 2y) = (10) \times 4 \quad \Rightarrow \quad 12x + 8y = 40$$
$$(-3) \times (4x + 5y) = (18) \times (-3) \quad \Rightarrow \quad -12x + -15y = -54$$

After multiplying the top equation by 4 and the bottom equation by –3, the two equivalent equations have a $12x$ and $-12x$ that will cancel out when the equations are added.

Practice 2

1. A system of equations has two linear equations that both have infinite coordinate pair solutions. What underlying technique is used in all three algebraic methods in order to solve a system of equations algebraically.

2. After solving for one of the variables, how do you solve for the second variable to complete your coordinate pair solution?

3. How do you check your solution to a system of linear equations?

4. When is *The Equal Values Method* the best method to use to solve a system of equations?

5. When is *The Substitution Method* the best method to use to solve a system of equations?

6. When is *The Elimination Method* the best method to use to solve a system of equations?

7. Solve the following system using the most efficient algebraic method: $y = -3x - 2$ and $y = 2x - 12$.

8. Solve the following system using the most efficient algebraic method: $y = 2x - 1$ and $-4x + 3y = -13$.

Answers

Practice 1

1. A system of equations is two linear equations being considered at the same time.
2. There are infinite coordinate pairs on any line or segment.
3. The coordinate pair where two lines intersect is the solution to a system of linear equations.
4. If a point (1,9) is a solution to a system of equations it means that (1,9) is a solution to both of the linear equations—when subbed into each equation it makes a true statement. (1,9) will also be the point on the graph where the two lines intersect.
5. The solution to the system is (4,5)

$y = -x + 9$

x	y
1	8
2	7
3	6

$y = x + 1$

x	y
1	2
2	3
3	4

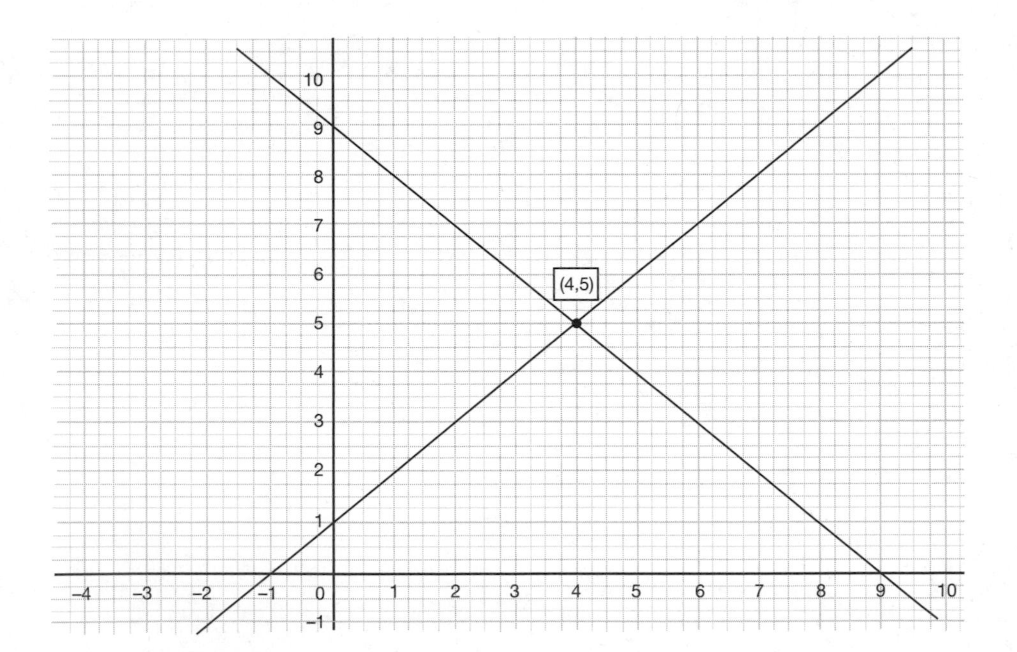

6. The solution to the system is (6,7)

$y = \frac{1}{2}x + 4$

x	y
2	5
4	6
8	8

$y = -\frac{1}{3}x + 9$

x	y
3	8
6	7
9	6

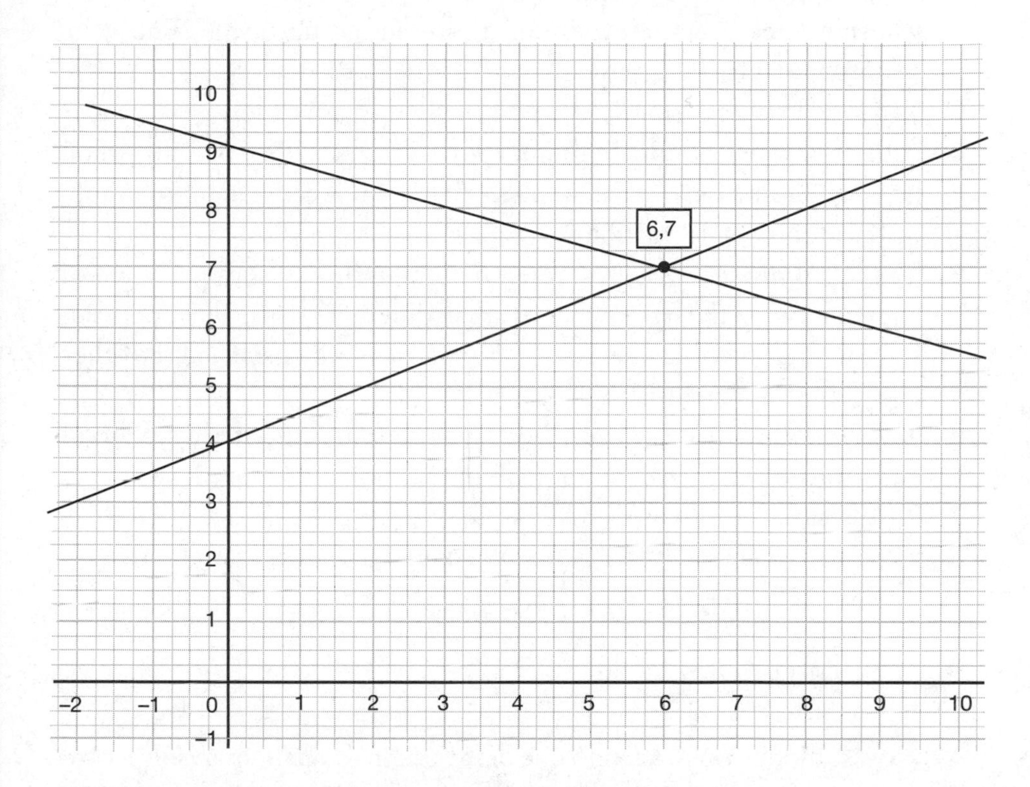

7. The solution (6,7) works in both of the following equations:

$y = \frac{1}{2}x + 4$; for the point (6,7)

$7 = \frac{1}{2}(6) + 4$

$7 = 3 + 4$ ✔

$y = -\frac{1}{3}x + 9$; for the point (6,7)

$7 = -\frac{1}{3}(6) + 9$

$7 = -2 + 9$ ✔

Practice 2

1. Each of the algebraic approaches relies on combining the two equations in a way that one of the variables is canceled out. That way, a single-variable equation remains and can be solved.
2. After solving for one of the variables, sub that value into either of the original equations to solve for the second variable.
3. A coordinate pair is a solution when it is a solution to both of the original equations. To check your answer, plug the coordinate pair solution back into the other equation you did not use to find your second variable.
4. *The Equal Values Method* is the best method to use to solve a system of equations when they are both in $y = mx + b$ format. It's especially useful if the equations have large numbers that are not convenient to graph, or if the solution is fractional.
5. *The Substitution Method* is the best method to use to solve a system of equations when the two equations are in different forms. This approach is particularly useful when one of the variables is by itself and the other equation is in the form $Ax + By = C$.
6. *The Elimination Method* is the best method to use to solve a system of equations when both equations are in the form $Ax + By = C$.
7. The best method to use for this system is *The Equal Values Method.*
   $y = -3x - 2$ and $y = 2x - 12$, so set the equations equal to each other:
   $-3x - 2 = 2x - 12$
   $-2 = 5x - 12$
   $10 = 5x$
   $x = 2$
   Now substitute this back into one of the original equations to solve for y.
   $y = -3x - 2$, for $x = 2$
   $y = -3(2) - 2$
   $y = -8$, so the solution is $(2,-8)$

8. The best method to use for this system is *The Substitution Method.* $y = 2x - 1$ and $-4x + 3y = -13$, so plug $2x - 1$ into the y of the second equation:

$$-4x + 3(2x - 1) = -13$$
$$-4x + 6x - 3 = -13$$
$$2x - 3 = -13$$
$$2x = -10$$
$$x = -5$$

Now substitute this back into one of the original equations to solve for y.

$y = 2x - 1$, for $x = -5$

$y = 2(-5) - 1$

$y = -11$, so the solution is $(-5, -11)$

25

Real-World Systems of Equations

STANDARD PREVIEW

In this lesson we will cover **Standard 8.EE.C.8.C**. You will learn when systems of linear equations have one solution, when they have no solution, and when they have infinite solutions. Then you will learn how to apply your skills with systems of linear equations to solve real-world problems.

The Three Cases of Systems of Equations

In the same way that single-variable equations can have one, infinite, or no solutions, systems of equations in two variables can also have one, infinite, or no solutions. In the previous lesson we covered systems of equations that have one solution and here we will discuss what the other two cases look like—on graphs and algebraically.

Systems of Equations with No Solutions

Remember Alex from Kokomo? Her mom wanted her to live at home for the summer and her dad wanted her to get a job. Those two demands had a solution. But what if her mom wanted her to live at home and her dad wanted her to stay at college for summer classes? She would *not* have been able to find a solution pleasing to both of them!

Similarly, sometimes two different linear equations have no solution in common. Lennart is thinking of two numbers that add up to 10. Yas is thinking of two numbers that add up to 100. There are lots of different numbers they could each have in mind, but could they both be thinking of the *same pair* of numbers? Is there a pair of numbers that can simultaneously add to 10 and 100? This situation can be modeled by the following system of equations:

$$x + y = 10$$
$$x + y = 100$$

When the side with all the variable terms is identical in both equations, it doesn't make sense that the other sides of the equations could equal different constant values at the same time. In these types of systems of equations there is always no solution.

There can also be no solutions when the variables have coefficients. For example, look at the following system of equations:

$$2x + 5y = 14$$
$$2x + 5y = -6$$

This system does not have a solution since $2x + 5y$ cannot equal both 14 and –6 at the same time for the shared (x,y) coordinate pair.

Lastly, keep your eyes out for equations in slope-intercept form that are identical except for their b value:

$$y = 3x + 10$$
$$y = 3x - 5$$

This is another example of two equations that do not have any (x,y) point in common, since tripling a number and adding ten will always produce a different answer than tripling the same number and subtracting five. We want to stress that sometimes some mathematical reasoning is all it takes to determine that a system of equations will not have a solution—you don't always have to move through all the algebraic steps to determine that!

Graphing Unsolvable Systems of Equations

Is your head hurting yet? Let's look at $y = \frac{2}{3}x + 1$ and $y = \frac{2}{3}x + 3$ on a coordinate plane to get a visual idea of *why* this is an unsolvable system of equations. In the following graph, the dotted line is $y = \frac{2}{3}x + 1$ and the solid line is $y = \frac{2}{3}x + 3$:

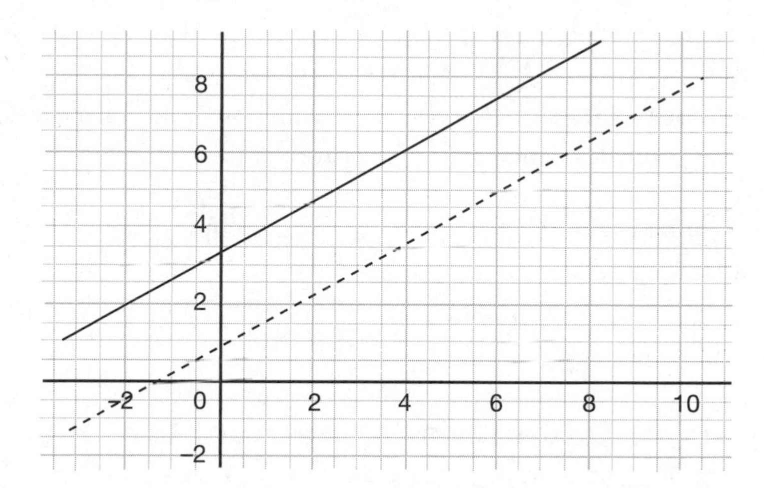

Notice that these two lines are moving up at the same rate and appear as if they will never touch. Also notice that they cross through the y-axis at different points. When two lines have the same slope and different y-intercepts they are called *parallel* and do not have a common solution.

Parallel lines in a coordinate plane will never touch. They have the same slope, but different y-intercepts. Parallel lines represent a system of equations that has no solution.

Practice 1

1. How can you tell that $3n + 2m = 5$ and $3n + 2m = 9$ will not have a common solution?

2. What will be true about how the two lines in question 1 relate to each other if they are graphed on the same coordinate plane? Where will they intersect?

3. Martino says that his two equations have no solution. If one of his equations is $y = \frac{3}{5}x - 1$, write two examples of what his other equation could be.

4. What are parallel lines?

5. What does it mean when a system of equations is graphed on the same coordinate plane and the lines are parallel?

6. Fill in the blanks to complete the sentence: When two lines are parallel in the same coordinate plane, they will always have the same ________________, but different ________________.

Systems of Equations in the Real World

Up until now, you may have thought that succeeding at math is just a way to get your high school degree, so that you can go to college, get a job, and ultimately move out of your parents' house and eat donuts and pizza at all hours of the day. While this is fairly good motivation on its own, we want to let you know that you should pay attention here, because mastering Systems of Equations will *also* help you determine what job is better for your personal needs, what company is better to make big purchases from, and what investments will bring you the most benefit. Viewing and solving real-world situations such as Systems of Equations really can help you make clever decisions—and who doesn't like being clever?

In Lesson 22, we met Kayla, who had a babysitting job. Her employers paid her $12/hour for babysitting plus an addition $10 for her transportation time. We determined that the equation $y = 12x + 10$ could be used to model her pay, where x = number of hours and y = total pay. Word gets out that Kayla has some incredible babysitting skills, so she starts getting some competing offers from other families for her summer months. She wants to decide what offer would earn her the most money, and she's eager to use her fancy new algebra skills, so she will compare these offers using Systems of Equations.

The Conlins versus the Youngs: An Algebraic Approach

The Conlin family is Kayla's current employer, who pay her according to the equation $y = 12x + 10$. The Young family approaches Kayla and say that while they don't pay a transportation fee, they will pay her $14 an hour, for as many hours a day she'd like to work. Kayla determines that the Young's pay scale can be represented as the equation $y = 14x$. She can't imagine working less than three hours a day, so she plugs three hours into each family's equation to see what each family would pay her for a three-hour day:

Conlin: $y = 12x + 10; x = 3$	Young: $y = 14x; x = 3$
Conlin: $y = 12(3) + 10$	Young: $y = 14(3)$
Conlin: $y = \$46$	Young: $y = \$42$
The Conlins pay $46 for three hours	The Youngs pay $42 for three hours

Kayla knows that since she won't have any school it's possible for her to work as much as eight hours a day. She plugs eight hours into each family's equation to see what each family would pay her for an eight-hour day:

Conlin: $y = 12x + 10; x = 8$	Young: $y = 14x; x = 8$
Conlin: $y = 12(8) + 10$	Young: $y = 14(8)$
Conlin: $y = \$106$	Young: $y = \$112$
The Conlins pay $106 for three hours	The Youngs pay $112 for eight hours

Kayla notices that the Youngs pay her more for an eight-hour day, but the Conlins' pay is greater for a three-hour day. She imagines that somewhere in between the three-hour shift and the eight-hour shift, there is a length of time where she would make the same amount of money from each family. She determines this amount of time by solving this system of equations algebraically. Since both the equations are in $y = mx + b$ form, we'll use the *Equal Values Method*:

$$y = 12x + 10 \text{ and } y = 14x$$

$$\begin{array}{rcl} 12x + 10 & = & 14x \\ \underline{-12x} & & \underline{-12x} \\ 10 & = & 2x \\ 5 & = & x \end{array}$$

So, at 5 hours, both families will pay the same

Plug this into one of the pay scale equations:

$y = 12x + 10; x = 5$

$y = 12(5) + 10 = \$70$

So, if she works five hours for either family, Kayla will make $70. Since the Young family pays $14 per hour, while the Conlin family pays just $12 per hour, Kayla realizes that if she plans on working more than five hours a day, she would earn more money from the Young family. Kayla decides that since it *is* her summer break, she probably doesn't want to work more than five hours per day on a regular basis. She decides to stay put with the Conlin family and the Young family is very disappointed.

The Conlins versus the Garcias: A Graphing Approach

A week after turning down the generous offer from the Youngs, Kayla gets another offer from the Garcia family. The Garcias pay $10/hour, but they give their babysitter $20 a day to use for transportation, food, or entertainment. If the babysitter is creative and doesn't mind cooking and entertaining the kids without spending money, that $20 is for the sitter to keep. When the Garcias ask Kayla if she might be interested in working for them she's not excited by the $10/hour pay, but she is curious how the extra $20 would help out. She decides to solve this system of equations through graphing. She represents their pay scale with the equation $y = 10x + 20$, makes two tables of points, and graphs the Conlin's pay scale with a solid line and the Garcia's pay scale with a dotted line:

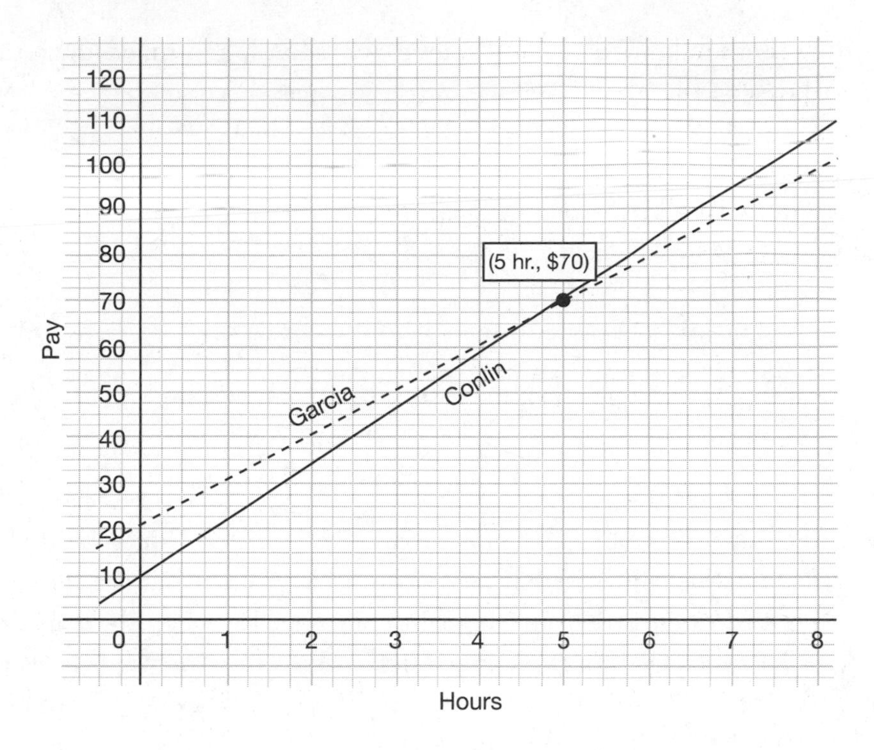

Kayla determines that the point of intersection of the system is at (5hr, \$70). She notices that when x is less than four hours, the line representing the Garcia family's pay scale is *above* the line representing the Conlin family's pay. This means that if she's working less than five hours per day, she'll earn more working for the Garcia family (assuming that she cooks and entertains the kids with her fascinating puppet skills so that she can keep the extra \$20 for herself). After (5hr,\$70) she sees that the Conlin's pay scale line is above the Garcia's, so she knows this means that for workdays longer than five hours, the Conlin Family will pay her more money.

The "Break-Even Point": Analyzing Solutions

Finding the "solution" to a system of equations doesn't always give us a solution to our real-world problem! That (x,y) coordinate pair solution is often referred to as the "break-even point." This is because both equations have the same y value for that specific x value, so the two equations are on an even playing field at that point. What you need to look closely at is what happens before and after that *break-even point*. One line will be steeper than the other and will be above the other line. Depending on the context of the question, this could symbolize that you'll be earning more money or paying more money for a product or service. You might want to take another look at the work Kayla did to really digest how these equations fit together.

It's sometimes helpful to plug x-values that are lower and higher than the x-coordinate of the *break-even point* into both equations to understand how they change. For now, let's take a look at Kayla's final decision!

All three families want Kayla to work part-time for them. They are tirelessly insistent! After analyzing all their pay scales, Kayla realized the following points:

- They will all pay $70 for a 5-hour work day.
- For days longer than five hours, the Young family pays the most.
- For days less than five hours, the Garcia family pays the most.

Since she is a savvy businesswoman (and a Systems of Equations Jedi), Kayla decides she will offer her babysitting skills to the Garcias for days she wants to work less than five hours, the Youngs for when she wants to work greater than five hours, and the Conlins for when she wants to work *exactly* five hours. This way she can keep all the families happy and not lose out on any potential earnings. Nice going, Kayla!

We hope you enjoyed watching Kayla analyze two different systems of equations so that she could make some great financial decisions for herself. She solved the first system of equations by an algebraic method and the second system by graphing. What was most important though was how she analyzed the different solutions she arrived at. Now we're going to look at one more example together where one of the equations is written in the $Ax + By = C$ form. This form is very common to use when translating word problems into algebraic models, so pay close attention!

To Raise Money or Awareness?

Katrina is starting a new program that helps children learn about growing food and caring for animals through volunteer time on a rural farm. She is planning an event to raise money for and awareness about this program, and she rented a space that can fit 200 people for an evening gala. Katrina plans to invite wealthy donors to attend for $100 with VIP tickets. The VIP tickets come with a professional photo of the donor with a newborn goat that has been named in honor of them. General admission tickets will be $20. A local restaurant has offered free catering, so everyone will enjoy dinner during a presentation on the program. This event is aimed at raising two important things that are vital to the success of this new venture: awareness and money! The more VIP tickets Katrina sells, the more money she'll make, but the less room there will be for the community members who will fuel this organization. Katrina determines that she only needs \$5,000 to get her program off the ground so she only wants to sell as many VIP tickets as necessary to meet that goal. She wants to know how many of the 200 tickets should be the \$100 VIP tickets and how many can be the \$20 general admission. Her friend Janchai helps her set up the following system of equations:

Step 1: Represent the total number of guests as a linear equation. Since there is only space for 200 people, Janchai determines that he'll let v = number of VIP guests and g = number of general admission guests. The equation $v + g = 200$ represents that there will be 200 guests.

Step 2: Represent the tickets sold and earnings goal as a linear equation.

Since VIP donors will pay \$100, $100v$ represents the money made from v number of VIP guests. Since general admission guests will pay \$20, $20g$ represents the money made from g number of general admission guests. Since Katrina wants the total income from both groups of guests to be \$5,000, Janchai will use the equation $100v + 20g = 5{,}000$.

Step 3: Set up the system of linear equations to be solved using elimination:

$$v + g = 200$$
$$100v + 20g = 5{,}000$$

Multiply both sides of the top equation by (–20) so that the g terms will cancel out when the equations are added together:

$$(-20)(v + g) = (200)(-20)$$
$$-20v + -20g = -4{,}000$$

Now add this equation to the other original equation:

$$\begin{array}{rcl} -20v + -20g & = & -4{,}000 \\ \underline{100v + 20g} & \underline{=} & \underline{5{,}000} \\ 80v & = & 1{,}000 \end{array}$$

$v = 12.5$, so approximately 13 of the tickets should be VIP tickets to raise \$5,000

Step 4: Plug $v = 12.5$ into one of the original equations and solve for the general admission tickets:

$$v + g = 200;\ v = 13$$
$$13 + g = 200$$
$$g = 187$$

So, 187 of the tickets should be general admission to spread the most awareness

Step 5: Check the coordinate pair (13,187) in the other linear equation:

$$100v + 20g = 5{,}000;\ (13{,}187)$$
$$100(13) + 20(187) = 5{,}000$$
$$5{,}040 > 5{,}000$$

As we can see, if Katrina sells 13 VIP tickets she will exceed her goal by $40, and she will get to host 187 general admission members. This solution will help her simultaneously meet her financial goal and inform the greatest number of community members about her program. She's ready to print tickets!

Practice 2

Use the following information to answer questions 1 through 4:

Wings Point restaurant pays servers $8/hour, and servers there average $38 in tips per shift. The Three Village Inn pays a flat fee of $80 per shift, plus $2 per hour. You have gotten job offers at both of these restaurants and have decided to compare these offers using a system of linear equations.

1. Write two different linear equations that model the pay rate for servers at each restaurant after *h* hours.

2. Looking at your two equations from question 1, which algebraic method would be the most efficient way to solve this system of equations?

3. Solve the system of equations and define what the break-even point coordinate pair means in real-world terms.

4. Analyze your results from question 3 and determine what would be the better job for you if you learned that the average shift length at both of these restaurants was six hours.

Use the following information to answer questions 5 through 8:

An apartment building in Wichita, Kansas, has a total of 50 apartments. There are studio apartments and one-bedroom apartments. Studio apartments rent for $425/month and one-bedroom apartments rent for $550/month. When all of the units in the apartment building are occupied, the total monthly rental income is $25,000. How many apartments of each type are there?

5. Let s = number of studio apartments and b = number of one-bedroom apartments. Write an equation that shows that when the studio apartments and one-bedroom apartments are combined, there are a total of 50 apartments in the building.

6. Let s = number of studio apartments and b = number of one-bedroom apartments. Write an equation modeling the information provided regarding the monthly rent for each type of apartment and the total rental income for the building. Hint: It will be in $Ax + By = C$ form.

7. Looking at your two equations from questions 5 and 6, which algebraic method would be the most efficient way to solve this system of equations?

8. Solve the system of equations and determine how many studio apartments and how many single-bedroom apartments there are in this apartment building.

Answers

Practice 1

1. It is not possible for $3n + 2m$ to equal five and nine at the same time. When two equations have two identical variable terms on one side of the equation, it does not make sense that the other sides of the equations could be different.
2. The two lines from question 1 will be parallel and they will never intersect.
3. $y = \frac{3}{5}x - 2$ or $y = \frac{3}{5}x + 1$. Any equation that is in the form $y = \frac{3}{5}x + b$ where b is any real number other than –1 is an equation that will have no solution in common with $y = \frac{3}{5}x - 1$.
4. Parallel lines are lines in a coordinate plane that have the same slope and never touch. They have different y-intercepts and they represent a system of equations that has no solution.
5. When two lines from a system of equations are parallel, it means the system of equations has no solution.
6. When two lines are parallel in the same coordinate plane, they will always have the same *slope*, but different *y-intercepts*.

Practice 2

Wings Point restaurant pays servers $8/hour, and servers there average $38 in tips per shift. The Three Village Inn pays a flat fee of $80 per shift, plus $2 per hour.

1. Let h = hours worked and p = total pay. Wings Point: $p = 8h + 38$. Three Village Inn: $p = 2h + 80$.
2. Since these are both in the $y = mx + b$ form, the *Equal Values Method* will be the most efficient way to solve this system.
3. $p = 8h + 38$ and $p = 2h + 80$, so set the equations equal to each other:

   $8h + 38 = 2h + 80$

   $6h + 38 = 80$

   $6h = 42$

   $h = 7$, now sub this into one of the original equations to solve for p

   $p = 8h + 38$, for $h = 7$

   $p = 8(7) + 38$

   $p = 94$, so the solution is (7,94)

   Since h is hours and p is pay, represent (7,94) as the break-even point of (7hr,$94). This means that if you worked seven hours at either restaurant you would earn $94 at both places.

4. Since we know that the pay will be the same at seven hours, let's look at the rate of change of each equation to determine what restaurant would pay more for six hours. Wings Point pays \$8 for each additional hour and Three Village Inn pays only \$2 for each additional hour. Therefore, it follow that shifts *longer than* seven hours earn more money at Wings Point since they have the larger hourly pay, but shifts less than seven hours would earn more at Three Village Inn, since they have the larger flat pay. We can plug six hours into each equation to verify our conclusion:

Wings Point: $p = 8h + 38$, $h = 6$. Wings Point pay would pay \$86.

Three Village Inn: $p = 2h + 80$, $h = 6$. Three Village Inn would pay \$92.

So if the average shift length is six hours, it's better to take the job at Three Village Inn.

5. $s + b = 50$

6. $425s$ = rental income from studio apartments and $550b$ = rental income from one-bedroom apartments. Since together the combined income is \$25,000, write $425s + 550b = 25{,}000$.

7. The system of equations $s + b = 50$ and $425s + 550b = 25{,}000$ will be most efficiently solved using the *Elimination Method*.

8. $s + b = 50$ and $425s + 550b = 25{,}000$.

Begin by multiplying the first equation by –425 so that the b terms will cancel out when the equations are added together:

$(-425)(s + b) = 50(-425)$ (distribute the –425)

$-425s + -425b = -21{,}250$

Now add this equation to the other original equation and solve for b:

$$\begin{array}{r} -425s + -425b = -21{,}250 \\ \underline{425s + 550b = 25{,}000} \\ 125b = 3{,}750 \end{array}$$

$b = 30$, so there are 30 one-bedroom apartments. Since there are 50 apartments all together, this means there must be 20 studio apartments.

26

The Different Faces of Functions

STANDARD PREVIEW

In this lesson we will cover **Standards 8.F.A.1** and **8.F.A.2**. You will learn what a function is and how to compare functions represented in equations, graphs, tables, and written descriptions.

What Are Functions?

A *function* is a mathematical expression where every x-value generates just one unique y-value. Where do we see functions in real life? Everywhere! Any equation that allows you to plug in an input to get a corresponding output is a function. For instance, determining the distance you've traveled using the equation $d = 50t$ is a function. How the babysitting pay you receive changes as your hours of work change is a function. Calculating a 20% tip on a bill is a function. Whenever we plug a value into an equation to get a different value out, we are working with functions. The most important thing is that each x-term has only *one* y-term. If you were paid $20 for two hours of babysitting for a family on Monday, but then received $30 from that same family for two hours of babysitting on Tuesday, that would *not* be a function since the same time (x-value) returned two different payments (y-values).

A function is a mathematical rule that assigns every x-value to exactly one y-value. Each input corresponds to just one output.

The Bake Sale Function

We can even look at a school bake sale as a loose example of a function! The type of item determines the price of the item. Since each type of item has just one predetermined price, that means it's a functional relationship. The bake sale volunteers wouldn't charge James $3 for an apple turnover and then charge Andrew $4 for the same apple turnover. That would be one messed up bake sale!

Mapping of Functions

Functions are often represented as mappings of points. The x-values are listed vertically first and arrows are drawn connecting them to their corresponding y-values. Here's what a bake sale price list might look like:

Glazed Donut	→	$0.50
Snicker Doodle	→	$0.50
Chocolate Chip Cookie	→	$0.50
Chocolate Brownie	→	$1.50
Blondie Brownie	→	$1.50
Apple Turnover	→	$2.50
Key Lime Pie	→	$3.50

When this is shown using function mapping notation, it would look like this:

Glazed Donut	→	\$0.50
Snicker Doodle	→	\$0.50
Chocolate Chip Cookie	→	\$0.50
Chocolate Brownie	→	\$1.50
Blondie Brownie	→	\$1.50
Apple Turnover	→	\$2.50
Key Lime Pie	→	\$3.50

Although each item is assigned to only one price, you might be wondering why some of the prices have more than one arrow pointing to them. The funny thing about functions is that while the x-values can have only one y-value, the y-values can have more than one x-value. When you get confused, just think about when you go into a store: there will probably be several items that are the same exact price, but the same exact item should not have different prices for different people!

Mapping of Non-Functions

An example of a real-world relationship that does not represent a function is the lottery tickets available for sale at a local charity event. People buy the tickets for an opportunity to draw tokens out of a \$1 bag, a \$5 bag, and a \$25 bag. Of course there are many more blank tokens in each bag than prize tokens:

- \$1 tickets give donors an opportunity to draw a blank token, a \$3 coffee voucher, or an \$8 ice cream voucher.
- \$5 tickets give donors the opportunity to draw a blank token, a \$30 gift certificate to a local restaurant, or a \$75 massage voucher.
- \$25 tickets give donors an opportunity to draw a blank token or a \$250 flight certificate redeemable on any airline.

As a mapping this would be modeled as such:

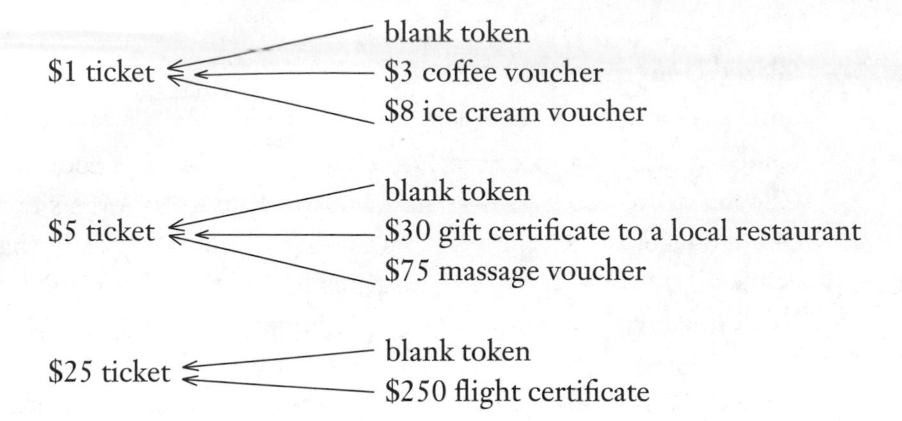

You can see from the mapping that since each input does not guarantee a single output, this relationship is not a function. Sometimes you will be asked to determine if a series of coordinate pairs represents a function. The above relationship would translate to the following pairs: ($1,0), ($1,$5), ($1,$8), ($5,0), ($5,30), ($5,75), ($25,0), and ($25,$250). As you may notice, when more than one coordinate pair has the same x-value and different y-values, a collection of coordinate pairs will not be a function.

Notice that a function does not have to come from a fancy mathematical equation. It does not have to look like a line on a graph and the points do not have to follow a pattern. As long as every *x*-value has one and only one *y*-value, a function can be any collection, or mapping, of coordinate pairs.

Linear Functions

Although a function does not *have* to be a linear equation, linear equations are the most common functions we will look at. A linear equation is a function since every x-value plugged into an equation will generate just a single y-value. If Jorge makes $10/hour at the record store and his boss always gives him $8 to cover his train fare, then his pay will be determined by the linear equation $p = 10h + 8$, where h = number of hours and p = pay in dollars. When talking about functions, the independent values are called the "inputs." Inputs are the x-values put into the function to determine the y-values, or "outputs." The coordinate pairs from a function can be put into a table of inputs and outputs:

Input (h)	Output (p)
1	$18
2	$28
3	$38

Remember that a linear equation will form a straight line on a coordinate plane. The following graph shows the relationship between Jorge's hours and pay. Notice that the coordinate pairs from the input-output table have been connected by a line. This line indicates that Jorge will get paid not only for whole number hours, but for partial hours as well:

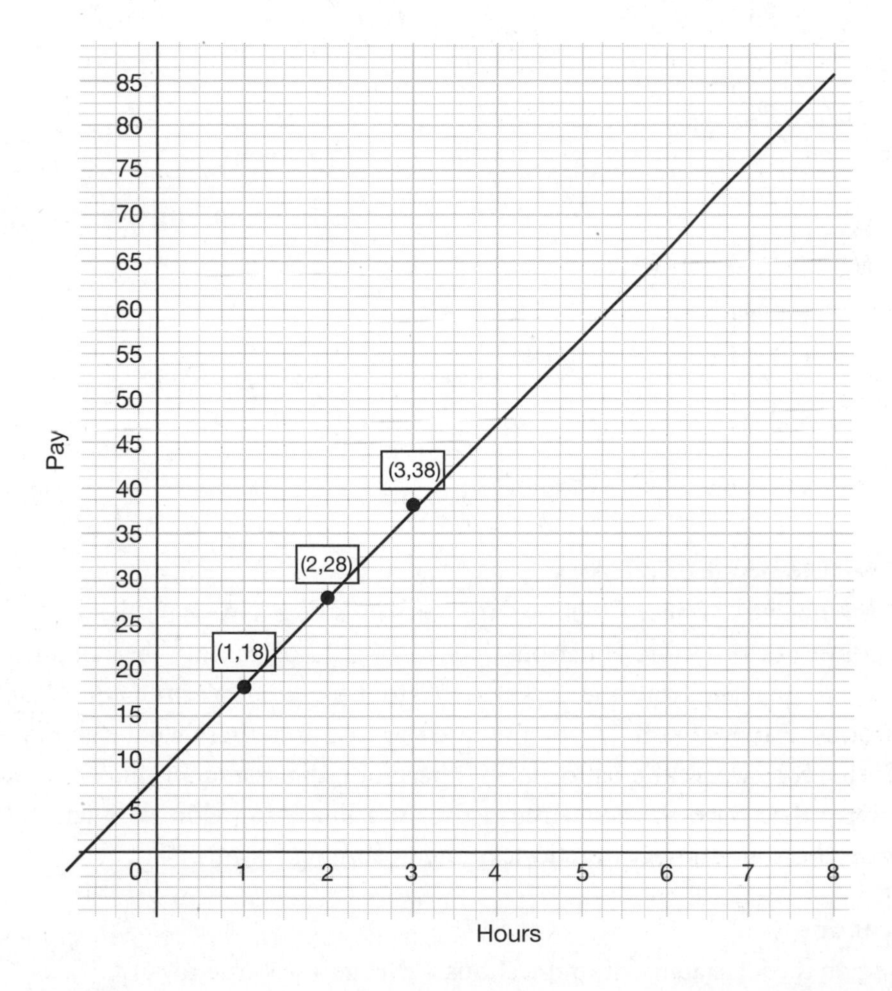

Sometimes it makes sense to *not* connect the dots when graphing a function! Suppose you are buying T-shirts as party favors for a birthday dinner you are hosting. They cost $10 and you will invite up to eight people. Since it is impossible to buy half a T-shirt, this relationship would be correctly graphed as a collection of individual points:

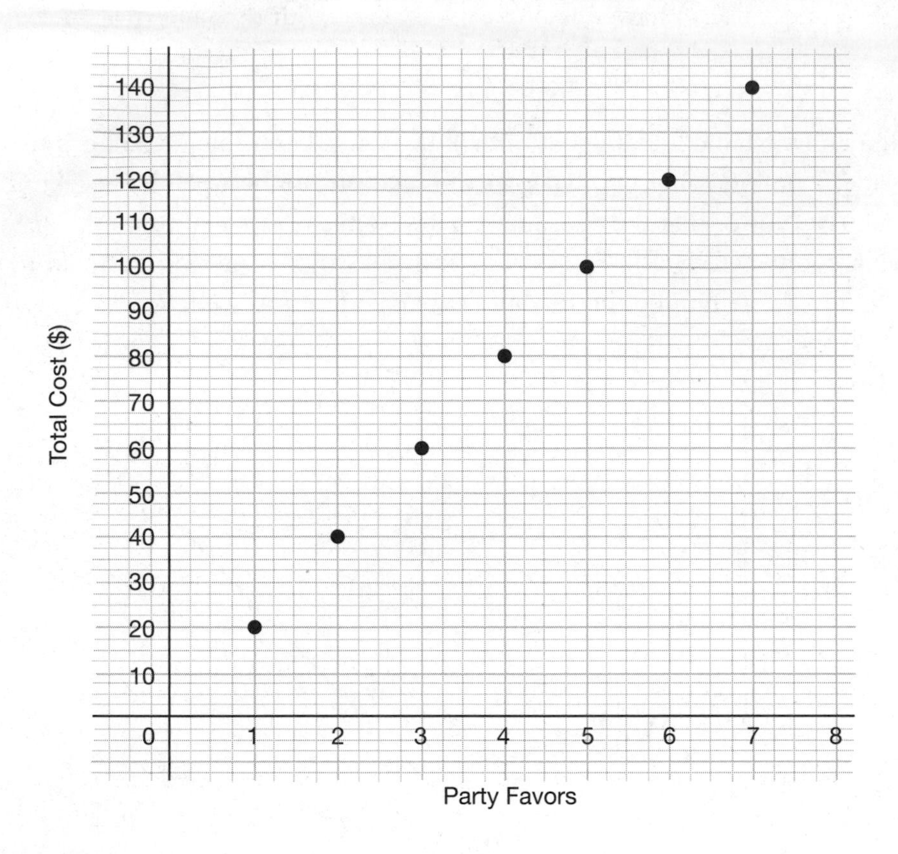

Non-Linear Functions

Remember that linear functions are equations that have variables with an exponent of one. The equation $h = -16t^2 + 200$ represents the height, h, of a ball dropped from the top of a 100-foot building over the course of t seconds. For every time, t, you put into the equation, there will be a unique height, h, of where the ball will be. Therefore, this is a function. Since the independent variable is squared, it is not a linear function. We will talk more about non-linear functions in the upcoming lessons!

Practice 1

Decide if each statement in questions 1 through 4 is true or false.

1. A function is always a line.

2. Any non-vertical line is always a function.

3. Functions can have two different x-values with the same y-value.

4. Functions can have two different y-values with the same x-value.

5. Which set(s) of points represent a function?

- **a.** (1,1), (2,2), (3,3), (4,4)
- **b.** (1,1), (2,1), (3,1), (4,1)
- **c.** (1,1), (1,2), (1,3), (1,4)
- **d.** (1,4), (–2,4), (3,4) (1,4)

6. If a function is defined by the rule $y = \frac{3}{5}x - 7$, what is the output when the input is 10?

7. If a function is defined by the rule $y = x^2 - 2$, what is the output when the input is 5? What is the output when the input is –5? Write your two input-output relationships as two coordinate pairs. Is this really a function? Explain your reasoning.

Comparing Functions in Different Formats

Like linear equations, function relationships are represented in many different ways. How often do you see an algebraic equation on your bill to double-check the tax you were charged? A function can be represented on a graph, as an equation, in a table, and in a written description. In order to be able to make sense of the world around us, we have to be able to compare functions presented in various formats. When given two functions in the same format, they can be easy to compare at times, but when they are in different formats it's important to be able to draw conclusions about rates of change, starting points, or specific values in each function. We will look at a few different examples of using the skills we have learned in the previous lessons to draw important conclusions when comparing different functions.

A Job Offer: Pay Scale Functions

Ethan is a code writer. He specializes in fixing bugs and adding new features to existing websites, and as a freelancer he's always looking for the next best contract. He has some family coming into town to visit over the next week, so he only has enough time to write about 1,000 lines of code before they arrive. He gets the following three offers and wants to determine which offer is best for his time constraints:

Offer 1: Ethan receives an email saying he would get an up-front signing bonus of $200 plus $0.50 for every line of code he submits.
Offer 2: Ethan gets a card of this table illustrating their pay scale:

Input—Lines of Code	Output—Pay
400	$200
600	$350
800	$525
1,000	$725
1,200	$950

Offer 3: A third company sends Ethan this graph illustrating what they pay:

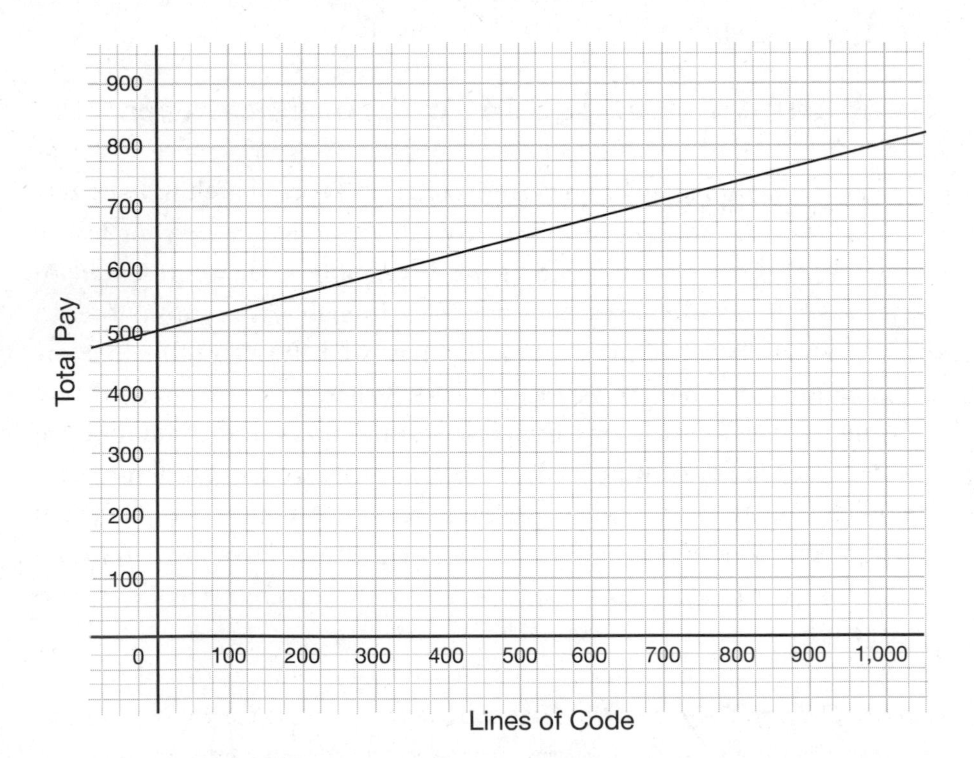

In order to decide between these three offers, Ethan needs to determine what he would earn with each contract if he completes the 1,000 lines of code he estimates he has time for:

> **Offer 1:** Since Ethan would receive a signing bonus of $200 plus $0.50 for every line of code, he concludes this is represented by the function $p = 200 + 0.50c$, where c = lines of code and p = pay. He plugs 1,000 in for c and calculates that he will earn $700 with this contract.
> **Offer 2:** Ethan looks at the table he receives in this second offer and sees that an input of 1,000 lines of code gives an output of $725, so he likes this offer better than the first offer since it is offering him $25 more compensation.
> **Offer 3:** For the third option Ethan takes a closer look at the graph. He locates 1,000 words on the x-axis, goes up until he hits the graph, and then moves horizontally to the left until he arrives at $800 on the y-axis. This is the best offer Ethan receives and he signs this contract.

Analyzing Functions: Rate versus End Result

Notice that in the example above we weren't just comparing the slope of each function. When functions have different starting points, it is critical to look at the whole picture and not just the *per unit* growth speed of a relationship. In the case with Ethan, it was important to evaluate each scenario to determine the exact pay Ethan would receive if Ethan met his goal of completing 1,000 lines of code. We were looking for the end result as a specific output based on a predicted input. You might remember that in an earlier lesson we analyzed another relationship where competing functions had different starting points. In Lesson 25, we didn't just focus on the rate of change (hourly pay) when we compared the babysitting offers of the Conlin, Young, and Garcia families. Recall that even though the Garcia family offered the least amount per hour (the lowest slope), the additional $20 they offered made them the most lucrative option for workdays less than five hours!

Although the starting point was a critical factor in Ethan's final pay, in the next example there are different starting points involved, but the real-world context demands that we ignore them and focus on the rates of change for each party. Why? We are going to be comparing rates (miles per hour) to see which duo cycled the fastest in their weekend training session. We'll need to use the slope formula, $m = \frac{y_2 - y_1}{x_2 - x_1}$, since these are not all proportional relationships and the starting points are not always at zero.

The Tour de Fargo: Speed Functions

The High Gear Cycling club is getting ready for the biggest race of the year: The Tour de Fargo. They normally train together on weekends but one weekend the team members are traveling, so they decide to work out in pairs and then report their training efforts to the team. The results are in, and the group scrambles to find out which pair of riders biked at the fastest average speed.

Pair 1: Victoria biked with Michael this weekend. She said her bike's mileage computer was at 2,348 miles when they started at 5:15 A.M. They finished their ride at 9:00 A.M., and at that time her computer read 2,430.5 miles.

Pair 2: Ali and Grey are gear junkies. They just got a new bike computer that sends data to their email accounts, which they can then print as a graph. They had already logged 200 miles on this new computer and forgot to reset it before their ride this weekend. They rode for 2.5 hours and their results can be read in this graph:

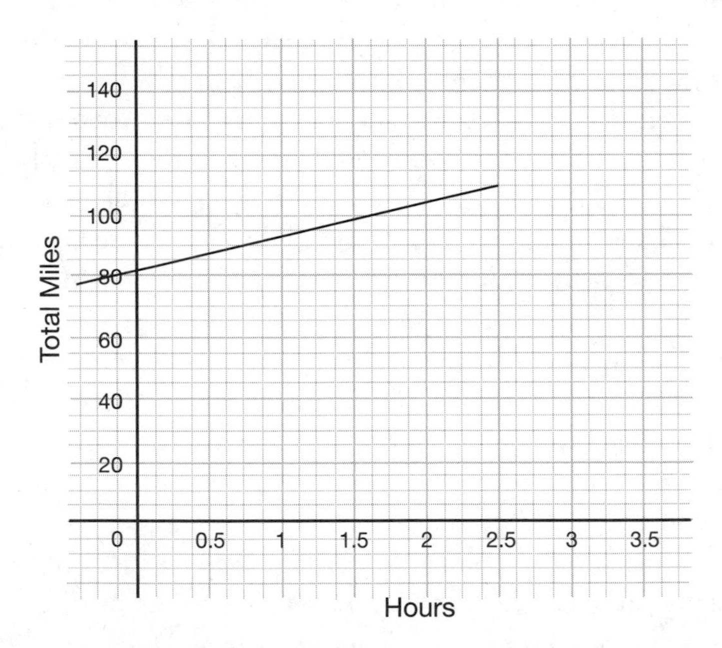

Pair 3: Chris and Jaap are the most reserved on the team but always bike the hardest. Without any pomp and circumstance they simply email the group, "we covered 91 miles in 260 minutes."

Time is the independent variable and distance depends on time. In order to calculate the average speeds of each dynamic duo, we are calculating how the distance cycled compares to the time ridden: $\frac{\text{change in distance}}{\text{change in time}}$. Let's go through each couple's reports:

Pair 1: Victoria and Michael started with 2,348 miles and ended with 2,430.5. Subtract these to see that they biked 82.5 miles. They started at 5:15 A.M. and finished at 9:00 A.M., so they rode for three hours and 45 minutes. Remember that we can't write 3.45 as their time because decimals are based on units of 10 and minutes are based on units of 60. (Also, if you think about it, 3.45 would be *less than* three and half hours and three hours and 45 minutes is almost 4 hours!) Change 45 minutes to a decimal by dividing it by 60: $\frac{45}{60} = \frac{3}{4}$, which is equivalent to 0.75. Since they biked 82.5 miles in 3.75 hours, put this information in the rate formula = $\frac{\text{change in distance}}{\text{change in time}} = \frac{82.5}{3.75} = 22$ miles per hour.

Pair 2: In order to find Ali and Grey's speed, look at their table to notice that in 2.5 hours their bike computer went from 200 miles to 250 miles. Since they bike 50 miles in 2.5 hours, their rate is $\frac{50}{2.5}$, which simplifies to 20 miles per hour.

Pair 3: To calculate Chris and Jaap's rate, let's use the formula *Distance = Rate* × *Time*, knowing that their distance was 91 miles. Since we need to calculate their rate as a *per hour* unit and not as a *per minute* unit, our first task is to change 260 minutes into hours. To determine how many 60-minute intervals go into 260, use division: $\frac{260}{60} = \frac{26}{6} = \frac{13}{3}$.

This is $4\frac{1}{3}$ hours, so we can immediately see that Chris and Jaap rode *the longest*, but let's put their distance and time into the formula to see how *fast* they rode:

$$Distance = Rate \times Time$$

$$91 \text{ miles} = \text{rate} \times \frac{13}{3} \text{ hr}$$

$$\div\frac{13}{3} \text{ hr} \qquad \div\frac{13}{3} \text{ hr}$$

$$\frac{91}{1} \times \frac{3}{13} = \text{rate}$$

21 miles per hour

So, the results are in!

- Victoria and Michael traveled at 22 mph
- Ali and Gray traveled at 20 mph
- Chris and Jaap traveled at 21 mph

We can conclude that Victoria and Michael traveled the fastest during their training ride. Since we don't know what kind of terrain and weather conditions each couple faced, we can't say with certainty that Victoria and Michael worked the hardest. For example, what if Ali and Grey were climbing steep hills, in the rain, at altitude in the Cascade Mountains? If that were the case, then their speed of 20 miles per hour may have meant a harder workout than at 21 or 22 miles per hour under different circumstances. What is most certainly fair to say though, is that every rider got some great exercise on their bikes this weekend! Good luck to them all at the Tour de Fargo!

Practice 2

1. Collette sells vacuum cleaners door to door. Her boss offers to pay her $50 for a 4-hour shift. Her boss hopes that Collette will sell at least two vacuum cleaners during her shift, but to give her additional incentive she has offered Collette an additional $15 for every vacuum cleaner she sells after two vacuum cleaners. Make an input-output table to represent Collette's pay when she sells anywhere from zero to six vacuum cleaners in an afternoon. Determine if this is a function.

Use the following information for questions 2 through 4:

Benicia has $1,865 in her bank account when she left to travel around Central America. After two weeks of using her debit card to fund her travels, Benicia has $1,219 left in her account. Her sister Tanisha just traveled for a few weeks to Asia and this graph shows her bank account balance during her trip.

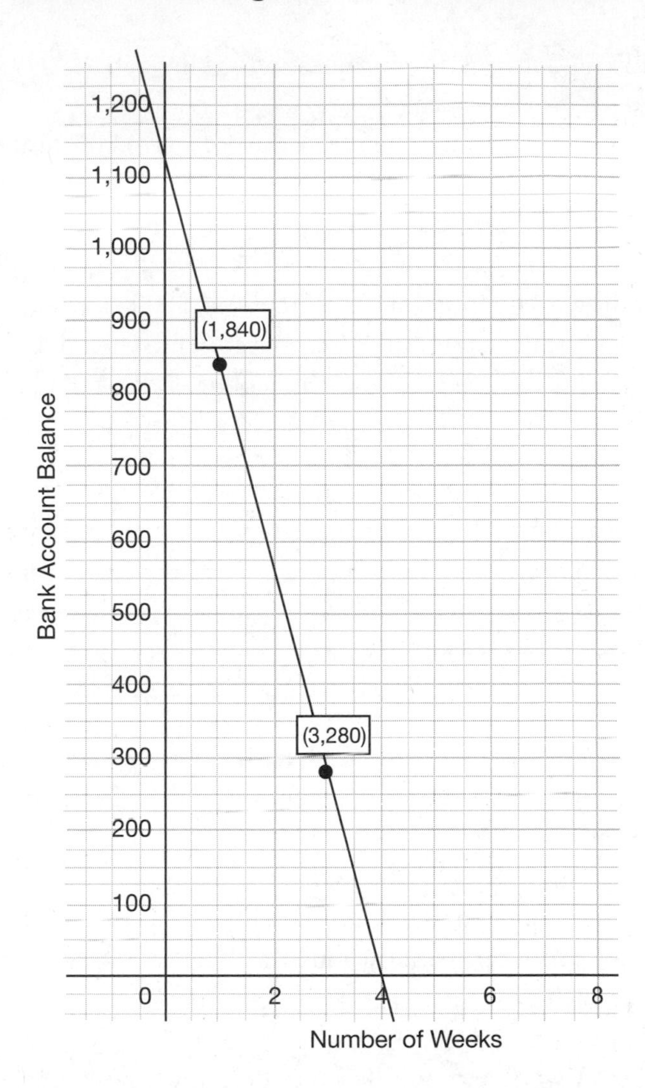

2. Approximately how much money did Tanisha begin her trip with? What was the balance in her bank account after four weeks?

3. What was Benicia's average spending per week? What function could be used to represent the balance in her bank account during her trip?

4. What was Tanisha's average spending per week? Which sister spent more money per week on her trip?

Use the following information for questions 5 through 9:

Parker is trying to determine where to invest a $20,000 inheritance she received from her great grandmother's estate. She goes to two different financial planners, Drew and Madeline, to see what they can offer her. Drew tells Parker that if she invests $20,000 with him, he will grow her money according to the function $P = \$20{,}000 + \$500(t)$, where t = number of years and P = the new amount of her investment. Madeline gives Parker the following table to illustrate how she will grow Parker's investment of $20,000 over the course of three, five, or eight years.

Time in Years	Amount of Investment
0	$20,000
3	$21,200
5	$22,000
8	$23,200

5. What is the rate of change that Drew is offering Parker? Explain this in real-world terms. How much will her investment grow with Drew each year?

6. What is the rate of change that Madeline is offering Parker? Explain this in real world terms. How much will her investment grow with Madeline each year?

7. Use your answer from question 6 to determine how much money Parker's investment will be worth after 10 years. (Hint: You know how much it is worth after eight years, according to the table.)

8. Use the formula to determine what the output will be if Parker opts for Drew's financial management and the input is 10.

9. Discuss what your answers to questions 7 and 8 mean when you compare them.

Answers

Practice 1

1. False—a function *can* be a line, but it can also be a collection of random points that do not make a line.
2. True—any non-vertical line is always a function. (A vertical line has just one x-coordinate with multiple y-coordinates, like (7,1), (7,2), and (7,5).)
3. True—Functions can have two different x-values with the same y-value.
4. False—Functions cannot have two different y-values with the same x-value. Every x-value must always have just one y-value.
5. a. Function: (1,1), (2,2), (3,3), (4,4)
   b. Function: (1,1), (2,1), (3,1), (4,1)
   c. Not a function: (1,1), (1,2), (1,3), (1,4)
   d. Function: (1,4), (–2,4), (3,4) (1,4) (The point (1,4) is repeated, but it does not violate the rule that every input needs to correspond with just one output.)
6. If a function is defined by the rule $y = \frac{3}{5}x - 7$, and the input is 10, the output will be $\frac{3}{5}(10) - 7 = -1$.
7. If a function is defined by the rule $y = x^2 - 2$, and the input is 5, the output will be $5^2 - 2 = 23$. When the input is –5, the output will be $(-5)^2 - 2 = 23$. These two coordinate pairs will be (5,25) and (–5,25), and this is still a function because each x value has only one y-value. In a function, two different x-coordinates can have the same y-coordinate, but the same x-coordinate input cannot generate two different y-coordinate outputs.

Practice 2

1. This is a function since every x-value returns just one y-value. The fact that she makes $50 whether she sells zero, one, or two vacuum cleaners does not keep this from being a function.

Number of Vacuum Cleaners Sold	Total Pay for 4-Hour Shift
0	$50
1	$50
2	$50
3	$65
4	$80
5	$95
6	$110

2. Tanisha began her trip with approximately $1,125. After 4 weeks Tanisha had no money left and her balance was at $0.
3. Benicia's average spending = $\frac{\text{change in } y}{\text{change in } x} = \frac{\$1{,}865 - \$1{,}219}{3 - 1} = \frac{\$646}{2 \text{ weeks}} = \$323$ per week. The function $b = 1{,}865 - 323w$, where w = weeks and b = balance represents the balance in her bank account during her trip.
4. Tanisha's average spending = $\frac{\text{change in } y}{\text{change in } x} = \frac{\$840 - \$280}{3 - 1} = \frac{\$560}{2 \text{ weeks}} = \$280$ per week. Even though Tanisha started out with much less money than Benicia, she was spending $21.50 less than Benicia each week.
5. The rate of change that Drew is offering Parker in the equation $P = \$20{,}000 + \$500(t)$ is $500 per year.
6. Use any two coordinate pairs to find the rate of change: (3, 21,200) and (8, 23,200): Rate of change = $\frac{\text{change in investment}}{\text{change in time}} = \frac{23{,}200 - 21{,}200}{8 - 3} = \frac{2{,}000}{5}$ The rate of change that Madeline is offering Parker in the equation $P = \$20{,}000 + \$500(t)$ is $400 per year.
7. Add $400 two times onto the 8th year amount of investment: $23,200 + $400 + $400 = $24,000
8. If Parker opts for Drew's financial management, the output will be $25,000 when the input is 10 years.
9. Since Parker would get an extra $100 per year through Drew's investments, over the course of 10 years her investment would grow an extra $1,000.

27

Linear and Non-Linear Functions

STANDARD PREVIEW

In this lesson we will cover **Standards 8.F.A.3** and **8.F.A.4**. You will learn how to distinguish between linear and non-linear functions. You will also learn how to construct a function to model a linear relationship represented in a graph

Linear versus Non-Linear Functions

You learned in the last lesson that a function is a mathematical relationship where every x-value has one and only one y-value. We focused on linear functions. We practiced recognizing functions and writing linear equations to model them. In this lesson we are going to define and deepen our understanding of linear functions as well as take a look at some important non-linear functions.

Linear Functions

When the points of a function form a line, the function is a **linear function**. A linear function is a linear equation where the value of each x input will influence the value of its corresponding y output. Linear functions cannot have any exponents other than 1, and every x will have one and only one y. Linear functions can always be rewritten in slope-intercept form, $y = mx + b$, but they can be presented in a variety of formats. These are all linear functions in disguise:

$$3x - 10y = 12$$

$$-\frac{7}{2} = y$$

$$\frac{1}{4}y - 8 = x$$

Each of these functions can be algebraically manipulated using opposite operations to fit into the model $y = mx + b$.

The Outcast Lines That Aren't Functions

The only straight lines on a coordinate plane that are *not* functions are vertical lines. Although vertical lines *are* linear equations, they are not functions because they violate the requirement that every x has one and only one y. Vertical lines are written in the form $x = k$, where k is any real number constant. Vertical lines are made up of infinite points that all have the same x-value paired up with different y-values. Notice that the linear equation $x = 5$ would produce the points (5,1), (5,2), and (5,10). Although these points connect to make a straight line, they do not satisfy the definition of a function.

Non-Linear Functions

If a collection of points doesn't make a straight line, it's not a linear function, but that doesn't mean it's not a function at all! There are many types of functions that are not straight lines and these all fall into the category of **non-linear functions.** Two common functions that we are going to discuss here are quadratics and absolute value functions.

Non-Linear Absolute Value Functions

If some friends tell you that they are five minutes away, you probably don't care if they are north, south, east, or west of your house—what is significant is that in five minutes they will be knocking on your door, most likely wanting some soda and chips. So, regardless of what direction they are from your home, you better get ready for them! Similarly, direction isn't important to absolute value. The absolute value of a number is the distance that number is from zero, regardless of whether the number is negative or positive. The absolute value function doesn't care about signs—it just reports back on the distance. The absolute value of a number is written by putting two bars on either side of it. We would read the expression $|8.5|$ as *the absolute value of* 8.5. Since 8.5 is 8.5 units from zero, we would write that $|8.5| = 8.5$. Similarly, –3 is three units from zero, so $|-3| = 3$.

Tables and Graphs of Absolute Value Functions

The most basic absolute value function is $y = |x|$. In this case, no matter what value you use as an input for x, y will be a positive number. Here is a table of some values of $y = |x|$:

x	y
-3	3
-2	2
-1	1
0	0
1	1
2	2
3	3

Notice that $y = |x|$ is a function since every x has one and only one y. However, $y = |x|$ is a *non-linear function* since it doesn't make a line. Lines continue straight in both directions and don't have bends or turning points.

Absolute value functions will always have the characteristic appearance of a "V" with a sharp turning point. Here's a graph of $y = |x|$:

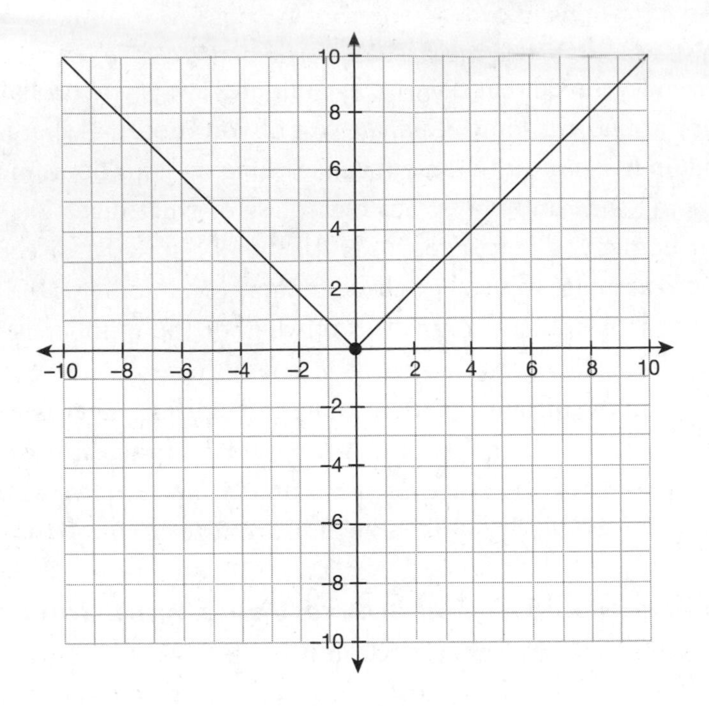

Absolute value functions come in all shapes and sizes. Here are some more examples of different types of non-linear absolute value functions that you might come across. It's important to remember that these are functions, they are just not linear functions!

$y = |\frac{2}{3}x|$
$y = |x - 4|$
$y = -|3x| + 4$

Non-Linear Quadratic Functions

We are looking into a crystal ball and we see that you will become intimately acquainted with *quadratic functions*. Actually, we don't need a crystal ball to know that—quadratic functions are really important functions, so of course you'll spend a lot of time on them in high school! A quadratic is a function that contains x^2. You might not think that sounds very important, but quadratics are used for modeling speed and trajectory, modeling business relationships in order to maximize profits or minimize costs, and to calculate complex speed measurements when wind speed or water speed adds an extra factor to a real-world situation. You're not going to dig too deeply into the nuts and bolts of quadratic functions until high school, but for now there are a handful of important things you should know:

- Quadratic equations can be written as $y = Ax^2 + Bx + C$ where A, B, and C are real numbers. In order to be a quadratic, $A \neq 0$ since if A was zero, the x^2 would cancel out and a linear function would remain.
- Quadratic equations are functions because every value of x produces one and only one y value.
- Quadratic equations are non-linear functions because they do not form a straight line on a graph. Instead, they form a U–like curve called a parabola. The U can face upward and look like a smile, or it can face downward and look like a frown.

Here is a table of some values that satisfy $y = x^2$ along with its graph:

x	y
-3	9
-2	4
-1	1
0	0
1	1
2	4
3	9

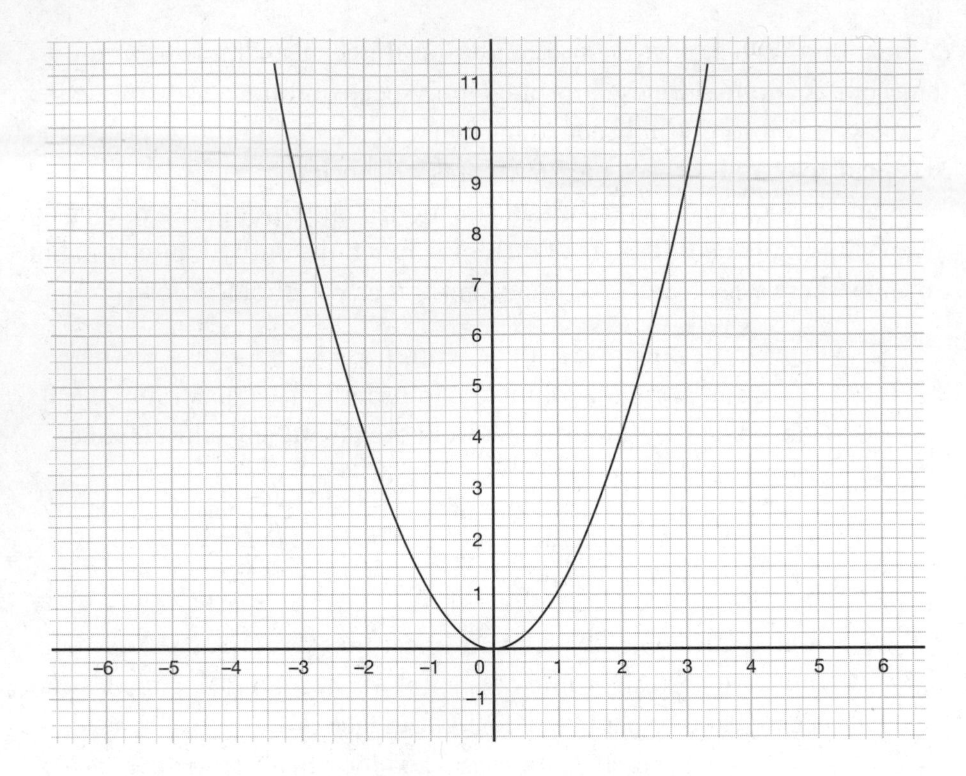

Notice that $y = x^2$ is a function since every x has one and only one y. It is similar to the $y = |x|$ function in that it has a turning point where it changes directions. It is different from the absolute value function in that its turning point looks more like a U and less like a pointed V.

Graphs of Functions

Not all mathematical relationships are functions. Sometimes you can tell this from looking at a table or a collection of points. Other times you will be able to quickly determine if a relationship is a function just by looking at the graph. For centuries (or at least decades), students have been using a foolproof method to determine if a graphed relationship is a function: the **vertical line test**. If a vertical line can be passed from the left edge of a graph to its right edge, without ever intersecting more than one point at any given time, then the relationship on the graph is a function. Conversely, if at any point a vertical line passing over a graph touches two points on a graph at the same time, then the graph is not a function.

Here is an example of the vertical line test in action. Notice that the vertical line passing through the graph on the left will never touch more than one point at any give time. The left graph passes the vertical line test and is a function. However, the vertical line passing through the graph on the right will touch multiple points on the graph at the same time. This graph fails the vertical line test and is not a function:

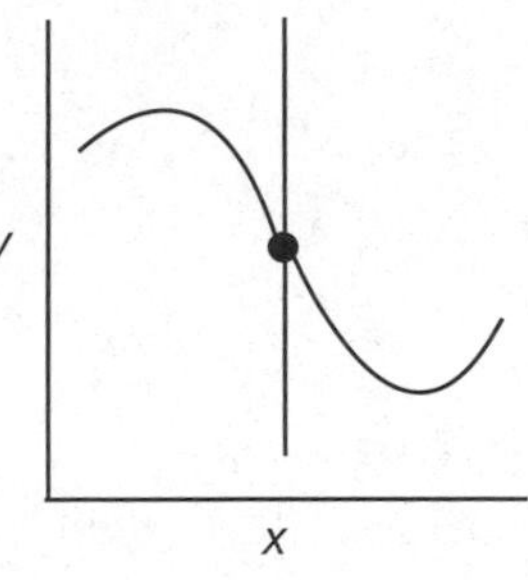

Function

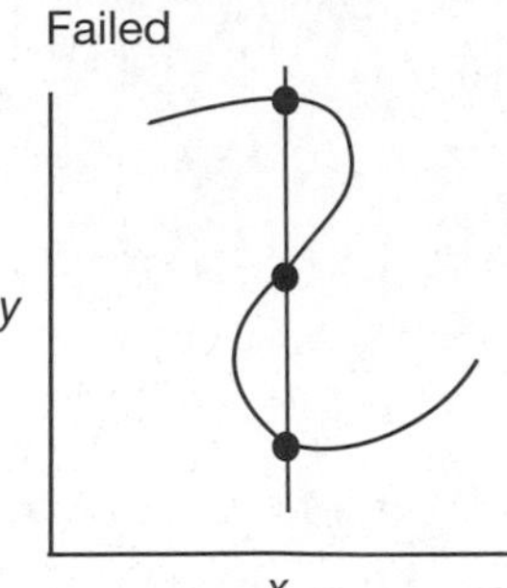

Not a Function

Practice 1

1. Is it true that as long as a line is straight it is a linear function?

2. Select all of the following that are functions:
   a. $y = |3x|$
   b. $x = 3$
   c. $y = 3x^2$
   d. $y = x^3$

3. Write a linear equation that is not a linear function. Explain why it is not a function.

4. What does the absolute value measure?

5. Make a table of values for $y = -|2x|$ and graph it.

6. Now make a table of values for $x = -|2y|$ and graph it. What do you notice about how the graphs from questions 5 and 6 compare? Which one represents a function and which one doesn't?

7. Select all of the following that are linear functions (you may want to graph some points with negative and positive x-values for the equations you are uncertain about).

a. $y = |3x|$

b. $x = 4y - \frac{1}{2}$

c. $y = \frac{1}{x}$

d. $-y = \frac{x}{7}$

e. $x = -5$

f. $y = 2x^2 + 2$

Digging Deeper into Linear Functions

Since all linear functions are also linear equations, all that we have learned about linear equations holds true for linear functions. We have practice creating graphs from tables or equations; determining the rate of change when given a table, graph, or pair of coordinates; calculating the starting point or y-intercept; and translating word problems into linear equations. In this section we will sum up all that we have covered and deepen our investigation of slope and linear functions.

Slope Formula

Let's review several of the most important things that we know about slope before moving on to some fresh concepts regarding slope:

- Slope is the *rate of change* between a dependent variable and an independent variable.
- The letter m is used to represent slope.
- For any points (x_1,y_1) and (x_2,y_2), $m = \frac{y_2 - y_1}{x_2 - x_1}$
- It's easy to mistakenly subtract your x-coordinates in the numerator, so care must be taken to make sure your y-coordinates go up top in the slope function.
- $\frac{\text{rise}}{\text{run}}$ can be used to map the relationship between any two points on a line with a slope triangle. The ratio of *rise over run* of any two slope triangles from the same line will reduce to the slope of that line.
- The slope-intercept form of a line is $y = mx + b$

One skill that has not been covered extensively is how to identify the slope of linear functions that are presented in formats other than the slope-intercept format, $y = mx + b$. When given a function in an alternate format, just use opposite operations to isolate y and rewrite the function so that it is in $y = mx + b$ form. Notice in the following example that when you divide as your last step to get the y alone, you must divide *both terms* on the opposite side of the equation.

Example: *What is the slope of the linear function,* $5x - 2y = 10$*?*
Solution:

$5x - 2y = 10$
$\underline{-5x \qquad -5x}$
$-2y = -5x + 10$
$\underline{\div -2 \quad \div -2}$ (you must divide everything by –2 because you are proportionally reducing everything in the original equation to half of its original value)

$y = \frac{-5}{-2}x + \frac{10}{-2}x$

$y = \frac{5}{2}x + -5$, so the slope of the linear function $5x - 2y = 10$ is $\frac{5}{2}$.

Positive, Negative, Zero, and Undefined Slope

The most important factor to consider when looking at the slope of a line is the *direction* the line is going. We read the direction of a line the same way we read a sentence—from left to right! The direction is so important because it communicates clear information about the relationship between the two variables—about whether x causes y to increase or decrease. In order to tell if a slope is positive, negative, zero, or undefined, imagine you are walking on the line from left to right. Here's a summary of the four different types of slopes to watch out for:

- When a line is pointing up, we say it is an *increasing function.* Increasing functions have positive slopes.

- When a line is pointing down, we say it is a *decreasing function.* Decreasing functions have negative slopes.

- When a line is flat, or horizontal, it does not have an increasing or decreasing slope. Flat lines have a slope of zero. (The $\frac{\text{rise}}{\text{run}}$ for horizontal lines is always $\frac{0}{\text{a number}}$, so their slope always equals zero.) Horizontal lines are written $y = k$, for any real number k, and are functions.

- When a line is vertical, it would be impossible to walk on! It therefore has an undefined slope. (The $\frac{\text{rise}}{\text{run}}$ for vertical lines is always $\frac{\text{a number}}{0}$, and since fractions are not allowed to have 0 in the denominator, their slope is always undefined.) Vertical lines are written $x = k$, for any real number k, and are *not* functions.

It is always important to identify how the independent variable, x, is influencing the dependent variable, y:

- In *increasing functions*, we say *as x is increasing, y is increasing*.
- In *decreasing functions*, we say *as x is increasing, y is decreasing*.
- In *horizontal functions*, we say *as x is increasing, y remains the same*.
- Vertical lines do not have an increasing x because x is always the same. Vertical lines are not functions.

Constant versus Changing Slope

Another important consideration of slope to keep in mind is whether the slope in a mathematical relationship is *constant* or whether the slope is *changing*. In straight lines the rate of change is always the same. If the rate of change of a linear function is $\frac{2}{3}$ at one part of the line, that means the slope will be $\frac{2}{3}$ for any part of the line. The rate of change is constant in lines: it does not speed up or slow down.

Linear functions are the only functions with a constant slope. All nonlinear functions have slopes that change. Quadratic functions have slopes that change as x increases. Let's look at an example so we can understand how that could be true. Imagine that a ball is dropped off the roof of a 50-story building. The speed of the ball represents that ball's rate of change. At the very first moment, the rate of change is zero because the ball is not moving yet. Once the ball is let go, the rate of change is a small number because the ball hasn't gathered any momentum yet and it is just falling slowly. However, as you can imagine, after a second or two, the ball begins to pick up more speed and starts tearing through the air. Its rate of change will increase to larger and larger numbers as its speed increases until it hits the ground. So in a case like this, as time in the air increases, the rate of change, or speed of the ball, also increases.

The y-Intercept: Where It All Begins

The y-intercept is the place where a function crosses the y-axis. It is commonly referred to as the *starting point* since it defines the value, or height, of the function when $x = 0$. Lots of real-world contexts don't have negative values for x, so the first value of x is often zero, and that represents where a function begins. An employee's base pay before commission, the beginning balance in a bank account before someone starts spending money on a vacation, and the length of a baby bear when it is born, for example.

The y-intercept is represented by the variable b in the slope-intercept form for linear equations, $y = mx + b$. You have had a considerable amount of practice calculating the slope between two points using $\frac{\text{rise}}{\text{run}}$ or $\frac{y_2 - y_1}{x_2 - x_1}$ and in Lesson 22 you learned how to algebraically find the starting point of a function. (We will cover it briefly here, but for a more in-depth explanation, please review Lesson 22.) Finding the starting point algebraically is important because sometimes the exact y-intercept might not be clear from a graph, or other times it would be inconvenient to make a graph of a large-scale relationship. In order to find the starting point, you must first calculate the slope, m. Then you plug m and one of the original (x,y) coordinate pairs into the $y = mx + b$ equations and solve for b.

Example: *A linear function passes through points* (2,14) *and* (6,8). *Find the starting point, interpret the slope, and write the equation containing these two points.*

Solution:

Step 1: Find the slope. $m = \frac{y_2 - y_1}{x_2 - x_1} = \frac{8 - 14}{6 - 2} = \frac{-6}{4} = -\frac{3}{2}$

Step 2: Find b by plugging $m = -\frac{3}{2}$ and (2,14) into $y = mx + b$:

$14 = -\frac{3}{2}(2) + b$

$14 = -3 + b$

$17 = b$

Step 3: Plug b and m back into the $y = mx + b$ equation: $y = -\frac{3}{2}x + 17$

Step 4: $y = -\frac{3}{2}x + 17$. The function has a starting point of 17 and a negative slope, so it is a decreasing function. For every two units that x increases, y decreases by three.

Writing Linear Equations from a Graph

In Lesson 22, you practiced finding the linear equation to model a relationship given three different cases:

> **Case 1:** You were given the slope and y-intercept.
> **Case 2:** You were given the starting point and one coordinate pair.
> **Case 3:** You were given two coordinate pairs.

You have actually learned how to write linear functions for the hardest types of questions already! Writing an equation from a graph is relatively easier! We'll add two more cases to the original list above and then show you how to handle these:

> **Case 4:** You will be given a graph where you can clearly identify the y-intercept and coordinate pairs on the line.
> **Case 5:** You will be given a graph where you cannot identify the fractional y-intercept but you can identify two coordinate pairs on the line.

Example (Case 4): *Find the equation of the following graphed linear function:*

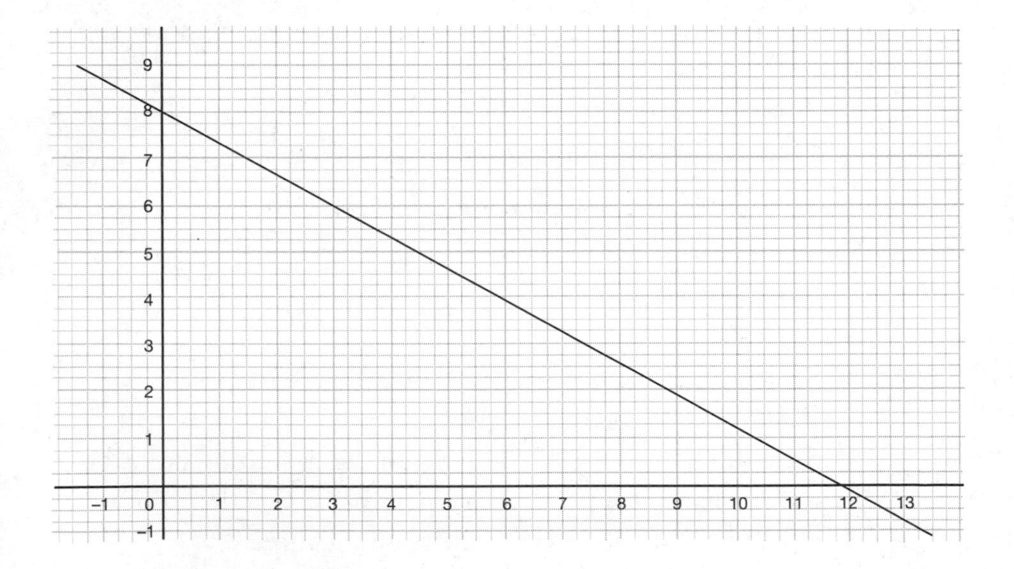

Solution: This is a **Case 4** type of question since we can identify that the *y*-intercept is at eight. We only need one other point since we can write the *y*-intercept as (0,8) and use that as one of our points, but in this case we will mark (3,6) and (9,2) on the graph. If we carefully select (*x*,*y*) coordinate pairs that have integer values, it is easy to calculate the slope by drawing in a slope triangle:

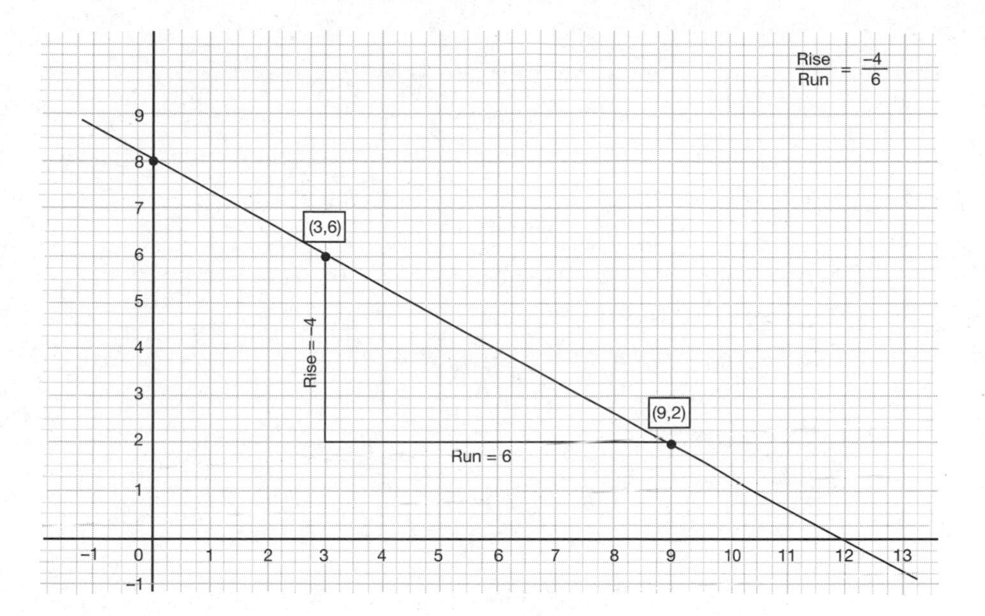

As you can see, the $\frac{\text{rise}}{\text{run}} = \frac{-4}{6}$, which reduces to $-\frac{2}{3}$. Therefore, our slope is $-\frac{2}{3}$ and we can see that our *y*-intercept is eight. We write the equation of the line as $y = -\frac{2}{3}x + 8$.

Example (Case 4): *Find the equation of the linear function graphed here:*

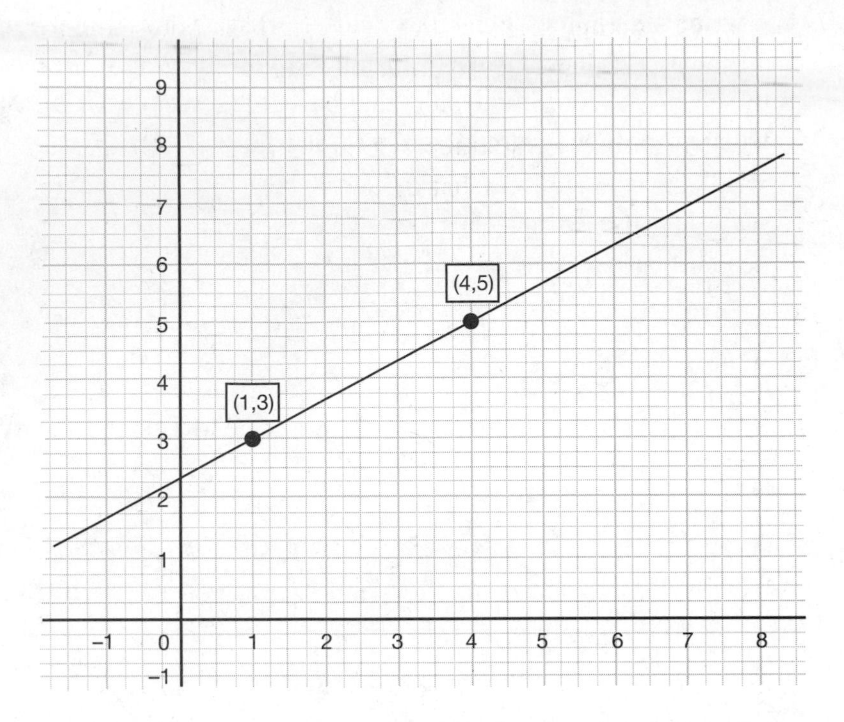

Solution: This is a **Case 5** type of question since we can identify that the y-intercept is somewhere between two and three, but we cannot say with certainty what its fractional value is. To solve a question like this, carefully search for two coordinate pairs that have whole numbers (we have selected (1,3) and (4,5)). Use these two coordinate pairs to determine that the slope is $\frac{2}{3}$ by using a slope triangle or using the slope formula $\frac{y_2 - y_1}{x_2 - x_1}$. Then plug $\frac{2}{3}$ and the coordinate pair (1,3) into the equation $y = mx + b$ to solve for the y-intercept. It is at $\frac{7}{3}$ and the equation for this line is $y = \frac{2}{3}x + \frac{7}{3}$.

Practice 2

1. Determine whether the following shapes would represent a function on a coordinate plane:

- **a.** a circle
- **b.** an upward facing parabola
- **c.** a sideways parabola that opens to the right
- **d.** The capital letter W
- **e.** The capital letter E
- **f.** $y = |xk|$, for any $k \neq 0$
- **g.** $x = |yk|$ for any $k \neq 0$

2. What are three different ways that slope can be identified in either a graph or an equation?

3. What is the difference between *a slope of* zero *and an undefined slope*? Explain what makes a slope zero versus what makes it undefined. What do lines with each of these slopes look like?

4. In what types of functions is the rate of change a constant and in what types of functions is the rate of change changing?

5. If the points (9,3) and (23,–3) are on the same line, what is the rate of change of that linear function?

6. Given the linear function $-3x + 4y = -8$, find the rate of change and the starting point.

7. If Gage goes into his winter season with 12 goals already scored, and then he scores five goals every three games, write a linear function to model how many goals he'll have, G, after n number of games.

8. Find the linear equation that models the function graphed here:

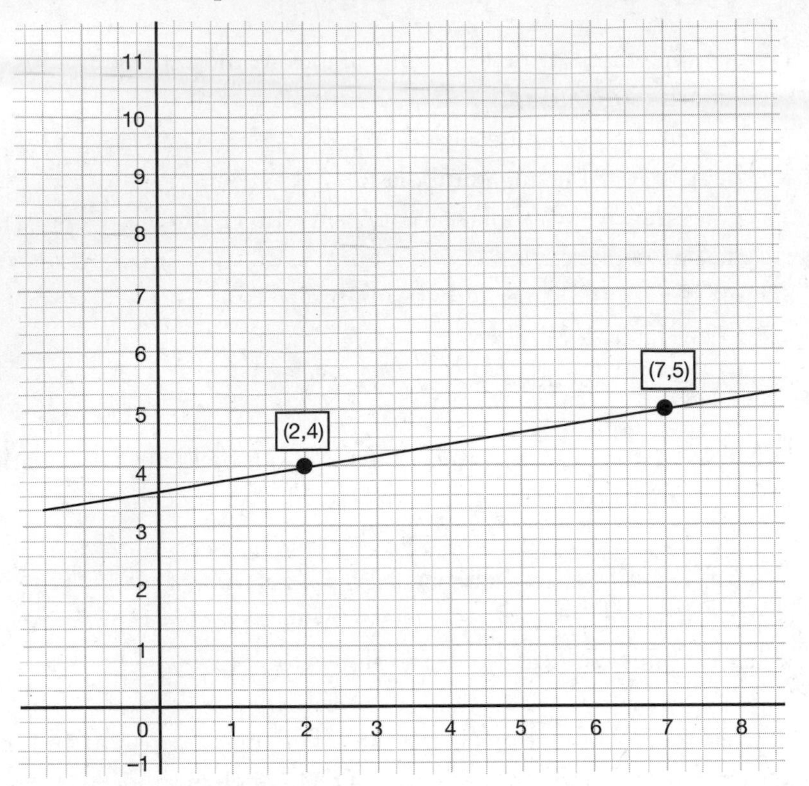

Answers

Practice 1

1. No—vertical lines are straight, but are not functions.
2. The functions include **a** $y = |3x|$; **c** $y = 3x^2$; and **d** $y = x^3$
3. $x = 1$ is not a function because it contains the points (1,0), (1,2), and (1,8), and therefore the same x value has multiple y-values and cannot be a function.
4. Absolute value measures how far a number is from zero without regard for whether the number is greater than zero (positive) or less than zero (negative).
5. $y = -|2x|$

x	y
-3	-6
-2	-4
-1	-2
0	0
1	-2
2	-4
3	-6

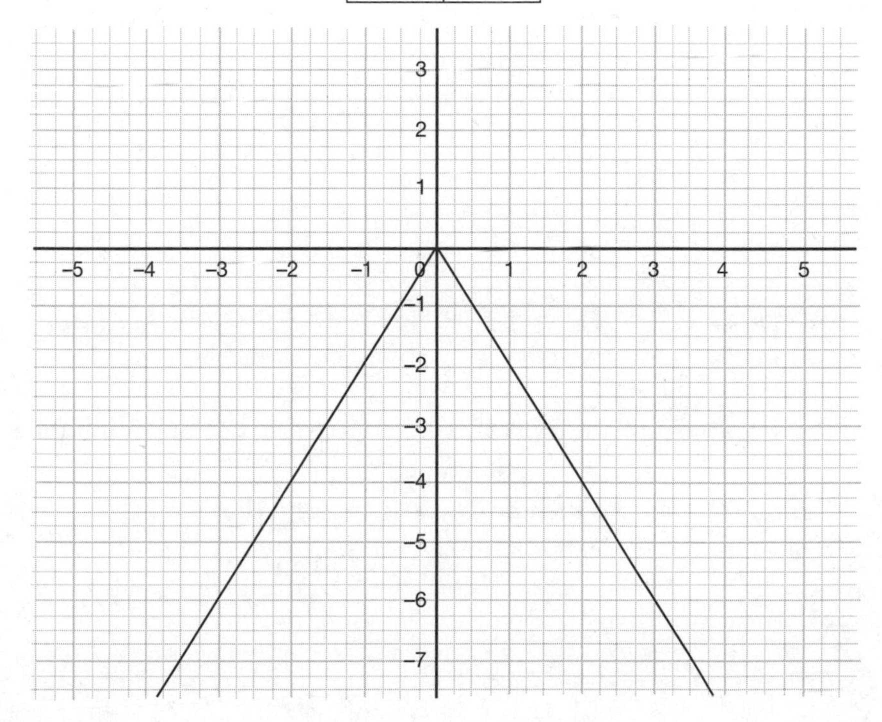

6. To make a table of values for $x = -|2y|$, plug in values for y and solve for x.

x	y
-6	3
-4	2
-2	1
0	0
-2	-1
-4	-2
-6	-3

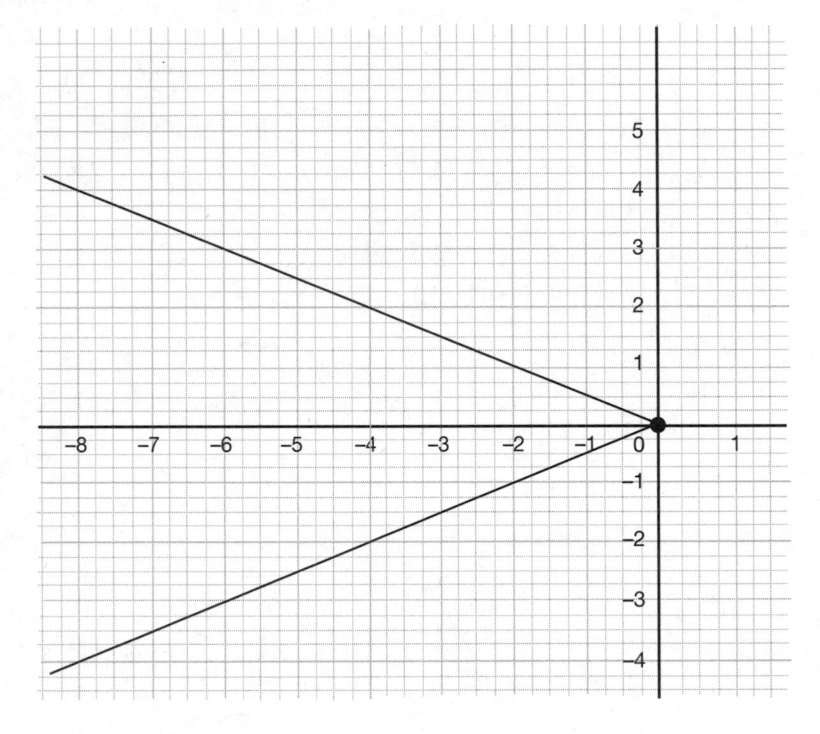

The graphs from questions 5 and 6 are similar, but the graph for $x = -|2y|$ is on its side and is not a function.

7. The linear functions include: **b** $x = 4y - \frac{1}{2}$ and **d** $-y = \frac{x}{7}$. Choice **e** $x = -5$ is a *linear equation*, but is not a *linear function*.

Practice 2

1. The following shapes would represent a function on a coordinate plane: **b** an upward facing parabola, **d** The capital letter W, and **f** $y = |xk|$, for any $k \neq 0$
2. The slope can bc identified with a $\frac{\text{rise}}{\text{run}}$ slope triangle, through the algebraic formula $\frac{y_2 - y_1}{x_2 - x_1}$, or through rewriting any linear function in the form $y = mx + b$, where m will be the slope.
3. A slope of zero has a rise of zero as the *run* increases. It is a horizontal line that does not slant up or down. An *undefined slope* has no run at all since x is the same value and only y is changing. This will look like a vertical line and these are never functions.
4. The rate of change is constant in all linear functions. The rate of change varies in all non-linear functions.
5. $\frac{-6}{14} = \frac{-3}{7}$. This rate of changes means that for every seven units that x increases, y decreases by three units.
6. The linear function $-3x + 4y = -8$ is equivalent to $y = \frac{3}{4}x - 2$ so the rate of change is $\frac{3}{4}$ and the starting point is -2.
7. 12 will be the starting point and the rate of change will be $\frac{5}{3}$. Therefore his goals after G games can be modeled $G = \frac{5}{3}n + 12$.
8. $y = 0.2x + 3.6$ or $y = \frac{1}{5}x + \frac{18}{5}$.

28

Analyzing Graphs and Describing Functions

STANDARD PREVIEW

In this lesson we will cover Standard 8.F.A.5. You will learn how to describe functional relationships through analyzing their graphs. You will also learn how to sketch graphs when giving a verbal description of the features of the function.

Describing Rate of Change

In the last lesson we learned that linear functions have a constant rate of change and all non-linear functions have a rate of change that changes. In this section we are going to look at how we can compare two different linear rates as well as two different non-linear rates. Since real-world information is often presented to us in graphs, being able to interpret how the relationship between two variables changes is a very important skill.

Comparing Linear Changes

We first investigated what different rates of change look like on graphs in Lesson 21 when Auggie, Jonah, Stella, and Remy were comparing what their parents paid them for helping out around the house. We learned that steeper lines represent greater rates of change. Lines that are more gradual represent slower change. In the previous lesson we learned that a horizontal line indicates that the rate of change is zero: although x is increasing, the value of y is remaining the same.

Analyzing Linear Slopes in Real-World Models

Let's pull together everything we've learned about slope to analyze what kind of tip revenue Karl had as a server one day at work. Karl arrives at the restaurant at noon and doesn't leave until 10:00 P.M., so the number on the x-axis represents the time of day:

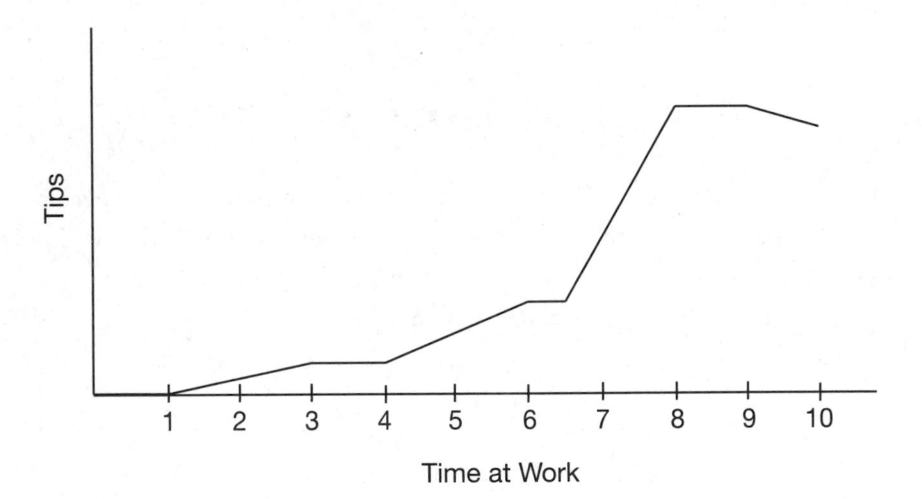

Notice that the hours have been marked along the x-axis, but there are no specific dollar amounts indicated along the y-axis. We don't need to know the specific dollar amounts in order to analyze the activity of Karl's day. We are going to use a table to organize our observations about how the slope changes throughout the day. Notice that in the first column we divide up Karl's shift into intervals of time that have common slope trends (increasing, flat, or decreasing). The second column is a summary of what the slope of the line is doing during that x-interval. Compare this table to the graph to make sure you agree with what it is presenting:

Time Intervals	Function Behavior
0 (noon) to 1:00 P.M.	Slope of 0
1:00 P.M. to 3:00 P.M.	Gradual increasing slope
3:00 P.M. to 4:00 P.M.	Slope of 0
4:00 P.M. to 6:00 P.M.	Steeper increasing slope
6:00 P.M. to 6:30 P.M.	Slope of 0
6:30 P.M. to 8:00 P.M.	Steepest increasing slope
8:00 P.M. to 9:00 P.M.	Slope of 0
9:00 P.M. to 10:00 P.M.	Gradual decreasing slope

Intelligently Defining Intervals

The word **interval** always refers to a range of x-values and is never used to discuss the y-values. When analyzing a graph, select intelligent x-intervals that contain similar slope activity. Part of analysis of real-world relationships is *reacting to the data* and not trying to fit it into a preconceived formula. Choosing regularly spaced x-intervals for this graph analysis would not make sense since the slope is not changing at regular intervals. Once you have broken up the data into intelligent x-intervals, describe what the slope is doing for each interval of x-values. According to our graph and table, we can see that there are four intervals where the slope is zero—this means that during these time frames, Karl's tip revenue was not increasing so he was not making any additional money. There are three intervals of increasing slope, which means that during those time frames Karl's tip revenue was increasing. Finally, there is one interval where the slope is decreasing, meaning that Karl's tip revenue decreased during that time frame. Let's add another column to our table to tell a story that could account for the slope variations in the graph:

Time Intervals	Function Behavior	Real-World Interpretation
0 (noon) to 1:00 P.M.	Slope of 0	Karl gets to work at noon—he does some prep work and begins waiting on his first diner of the day.
1:00 P.M. to 3:00 P.M.	Gradual increasing slope	Lunch is slow at this restaurant but Karl starts getting some tips during these two hours. His rate of tip revenue is low, but positive.
3:00 P.M. to 4:00 P.M.	Slope of 0	The restaurant closes from 3:00–4:00. The servers eat an early dinner and get prepared for the dinner service.

4:00 P.M. to 6:00 P.M.	Steeper increasing slope	This restaurant is known for its early dinner specials, and they are packed—Karl's tips increase nicely for two hours.
6:00 P.M. to 6:30 P.M.	Slope of 0	Karl gets a 30 minute break between tables and makes no tip revenue.
6:30 P.M. to 8:00 P.M.	Steepest increasing slope	Karl gets a table of 10 with one of his highest-tipping regular customers—his tips increase at the highest rate of the night during this time.
8:00 P.M. to 9:00 P.M.	Slope of 0	The restaurant empties out and Karl has to do clean-up work for an hour, during which he makes no tip revenue.
9:00 P.M. to 10:00 P.M.	Gradual decreasing slope	Karl tips out the busperson and the kitchen staff before going home for the night. It was their hard work that helped him make good tips, so his total tip revenue decreases as he pays them for their help.

Negative and Flat Slopes

One of the most difficult things for students to grasp when analyzing graphs is how to interpret decreasing and flat slopes. Notice that when there is a slope of zero, it does not indicate that the *x-value* stops in time—that continues regardless of the slope. What happens when the slope flattens out is that the y-value has ceased increasing or decreasing and is remaining steady. So in this case, *time* is moving forward, but *tips* are remaining steady. Similarly, a negative slope is not an indication that x is decreasing, but it is instead an indication that y is decreasing.

ERROR ALERT! Sometimes it is easy for students to make incorrect conclusions about other factors being related to a negative slope. For example, on a graph that compares distance over time, a negative slope might trick some student into thinking that a body in motion is *slowing down*; however, the negative slope is really an indication that the *distance* is decreasing.

Interpreting y- and x-Intercepts

You are already very familiar with the y-intercept, which is the starting point or initial value of a relationship. Since the y-intercept is the point where $x = 0$, in a functional relationship there will only be one y-intercept since when x is zero, there can only be one and only one value for y. However, the same is not the case for x-intercepts. An x-intercept is an indication that a specific input of x has created an output of zero. All x-intercepts are points where $y = 0$ and they are therefore commonly referred to as *break-even points*. Break even refers to the fact that the value of the function is not positive or negative at an x-intercept, but is instead equal to zero. Look at the following graph that represents the bank balance of a retail store over the course of a year:

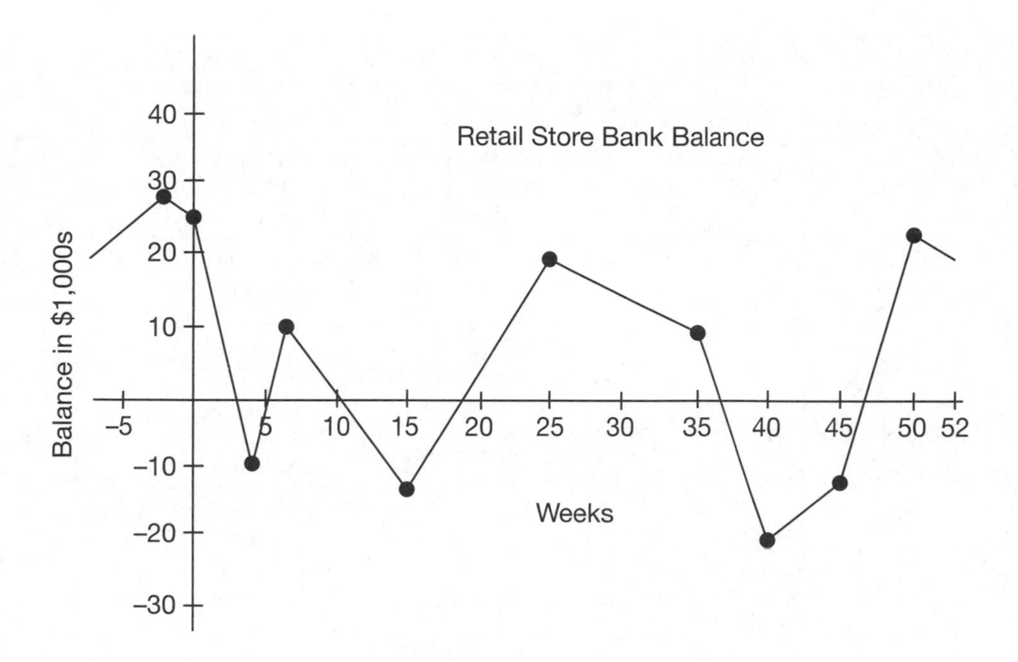

In this graph, the x-axis is divided up by weeks. The y-axis represents the start of the new year and the y-intercept represents the initial balance the store starts the year with, which is around $25k. The negative x-values represent the final weeks of the previous year. We can see that during the final weeks of the previous year, the balance was increasing, but during the last week of the year the balance dropped. Perhaps people were returning unwanted holiday gifts!

Starting at the y-intercept and moving forward, it looks like the store had to do a heavy amount of purchasing of goods for the new season, or maybe it needed to pay for other major expenses. We see a steep decreasing slope where the bank balance went down to almost negative $10k. Around three weeks we see that the store's bank account balance was at zero. The store didn't have extra money in the bank, but they weren't in debt. Each time this function crosses through the x-axis, it is representing that the store is neither making money nor losing money at that exact point in time. Looking at the trends in this graph it looks like Valentine's Day in February, Mother's Day in May, and the winter holidays are the busiest times of the year. Hopefully next year they will have more of their graph *above* the x-axis and less serious dips into debt!

Practice 1

The following graph represents Sierra's distance from home over the course of a 90-minute training run.

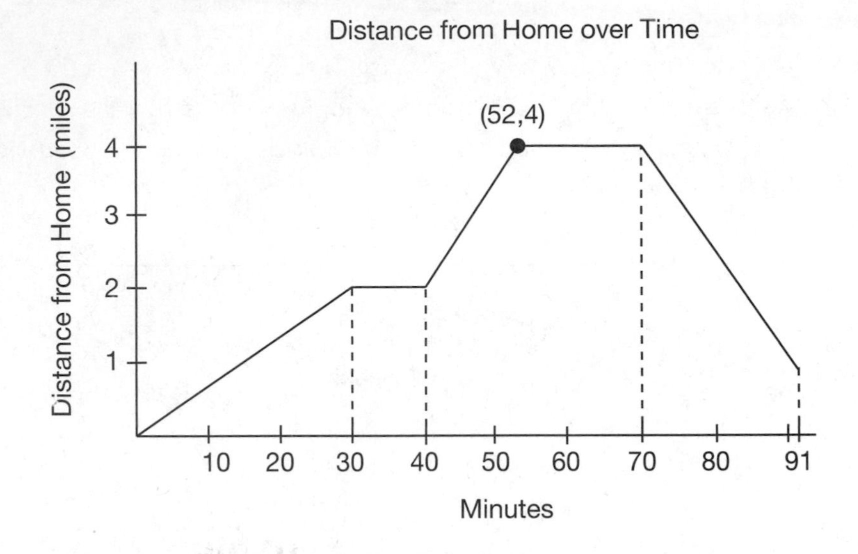

x-Intervals	Function Behavior	

1. Determine the best intervals to use for analyzing the function. Fill these intervals into the first column of the table. Then describe what behavior the function is exhibiting in the second column. Leave the third column blank.

2. What is the rate of change in simplest terms during her first two miles? Describe what this means in real-world context.

3. What do you notice about the interval from 30 to 40 minutes? What could account for this?

4. How does the slope of the line during the interval of 40 minutes to 52 minutes compare to the slope of the line from zero to 30 minutes? Calculate Sierra's rate of change in simplest terms and compare it to the rate found in question 2.

5. What does the negative slope indicate during the interval of 70 minutes to 91 minutes? Is Sienna slowing down? Describe her rate of change during this time frame in real-world context and compare it to her other rates.

6. In total, how many miles did Sierra run? Where did her run end?

7. Now label the third column "Real-World Interpretation" from the previous table to tell a story that could be an accurate description of Sierra's run.

Describing Non-Linear Relationships

In the last section we learned how to make sense of mathematical models that were comprised of sections of straight lines with positive, negative, and zero slopes. In this section we are going to learn how to read and describe non-linear graphs.

Changing Rates of Change

Many situations in the real world do not have steady rates of change and instead have rates of change that are always in flux. Think about what happens when you're in a car entering a highway. Once the traffic light on the entrance ramp turns green, it's important to speed up with an increasing rate of change so that the car can quickly get up to 40 to 50 mph so that it is safe to enter the highway. It is important for that driver to increase their speed at an increasing rate. Once the driver reaches around 50 mph, they probably release the pressure on the gas pedal a little bit so that the rate of increase begins to slow down—they want to carefully approach 65 mph so that they don't end up racing up to 80 mph and getting a ticket. At this point, the car's speed is *still increasing*, but at a *decreasing* rate of change. We know this might be making your head hurt a bit, so let's focus on rates of increase.

The Three Types of Increase

There are three different ways that a function can increase:

1. A function can increase at an increasing rate
2. A function can increase at a steady rate
3. A function can increase at a decreasing rate

You already know that a steady rate of increase will produce a linear function, so we will only look at the first and third types of functions listed. Since the steepness of a function is an indicator of how great the rate of change is, we can observe how the steepness changes in the functions in order to analyze them. (Remember that the word *interval* is used to indicate ranges for the values of x.)

Increasing Rates of Increase

The following graph is an example of a function with an increasing rate of increase. The slope of any part of the function is positive, so this is an increasing function. However, the rate of increase is increasing as x gets larger. We'll break the x intervals down into sections of two units. Look at the notes under each section of the graph and then read the description that follows in the paragraph after the graph:

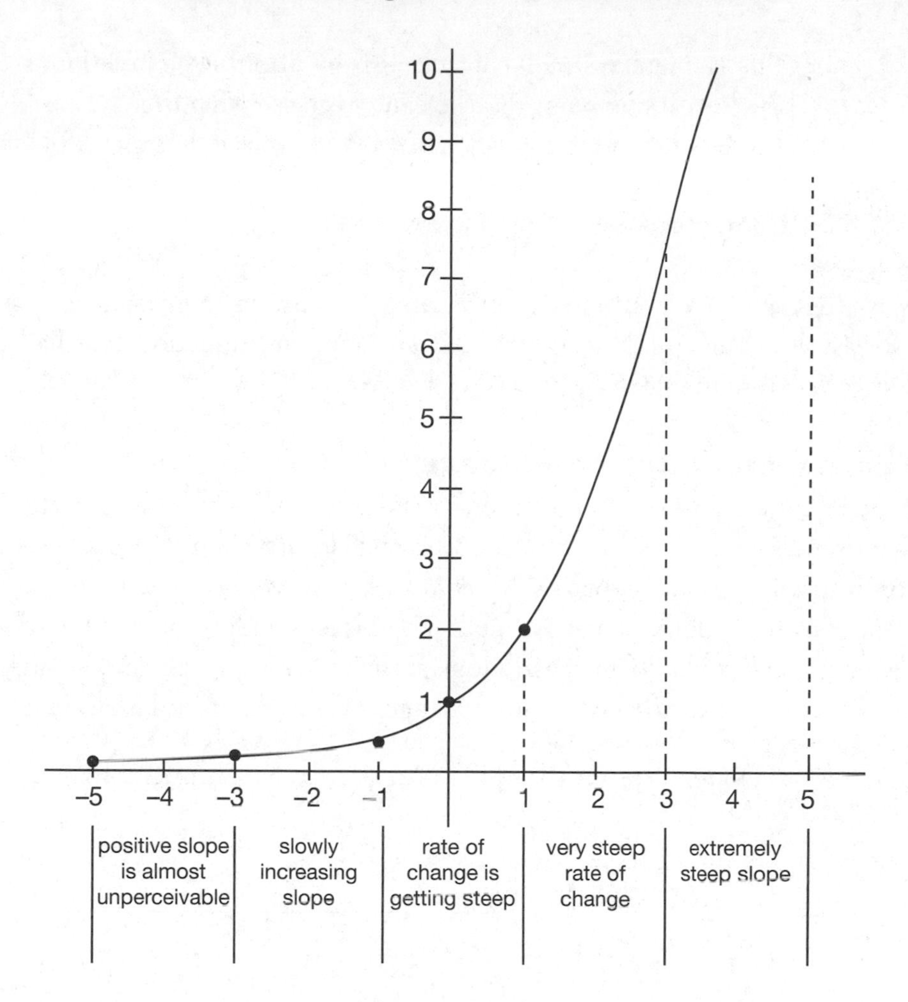

- On the interval of –5 to –3 the slope is relatively flat; the rate of increase is almost unperceivable.
- On the interval of –3 to –1, the rate of change becomes more recognizable, but is still a very gradual slope. (Notice that y increases by less than $\frac{1}{2}$ in this interval)
- On the interval of –1 to 1, the steepness is starting to grow. (Notice that y increased by about 1.5 in this interval, so this slope is about three times as steep as the previous interval's slope.)
- On the interval of 1 to 3, the function is becoming very steep. (Notice that y increased by six units in this interval, so it is about four times as steep as it was in the previous interval.)

- The last interval we'll consider is from 3 to 5. The function is so steep at this point that we can't even see what the y-value will be of the function when $x = 5$. We can see that the slope of this function is growing the most rapidly here from the extreme steepness of the function.

Now that we have examined what a function with an increasing rate of increase looks like, let's take a look at an increasing function that has a decreasing rate of increase.

Decreasing Rates of Increase

When a function's slope gets steeper, or closer to vertical, it is an indication that its rate of increase is *increasing*. Next we will look at a function that is also continuously increasing, but its *rate of increase is decreasing*. When you see a slope start to mellow out and get more flat, this is an indication that the rate of change is slowing down. Take a look at the following graph, and identify what you notice about general slope changes over each interval indicated:

- As x goes from –5 to –4 ______________________________
- As x goes from –4 to –2 ______________________________
- As x goes from –2 to 1 ______________________________
- As x goes from 1 to 7 ______________________________

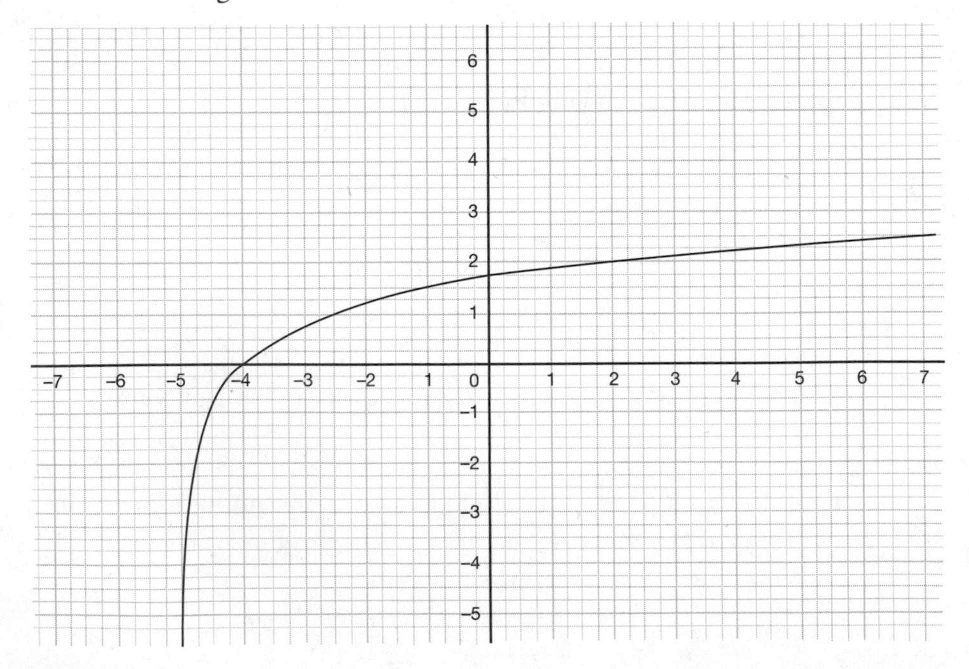

Hopefully after looking at the graph you were able to conclude that the rate of change is incredibly high on the interval of –5 to –4 and that over the interval of –4 to –2 it is still steep but starting to slow down. After –2, the rate of increase has slowed more, and as x gets larger than 3, the slope of the functions continues to become less steep. The rate of increase is decreasing as x gets bigger. Next we will investigate how rates of decrease can vary.

Varying Rates of Decrease

Rates of decrease can change at decreasing or increasing rates. If the driver realizes her exit off the freeway is approaching in two miles, she will begin to slow down moderately. But if something large and dangerous blows into the highway ahead, she will decrease the speed of the car at an *increasing rate*. Now that you have some experience being a slope detective, take a look at the following two graphs that illustrate decreasing functions. Which function has a decreasing rate of decrease and which one has an increasing rate of decrease? Read the slopes from left to right: as a slope flattens out it is showing a decreasing rate of change. As slope gets steeper it is an indication of an increasing rate of change. Organize your thoughts about these two graphs and then check your conclusions in the paragraph below the graphs.

Which function is decreasing at an increasing rate of change?

Which function is decreasing at a decreasing rate of change?

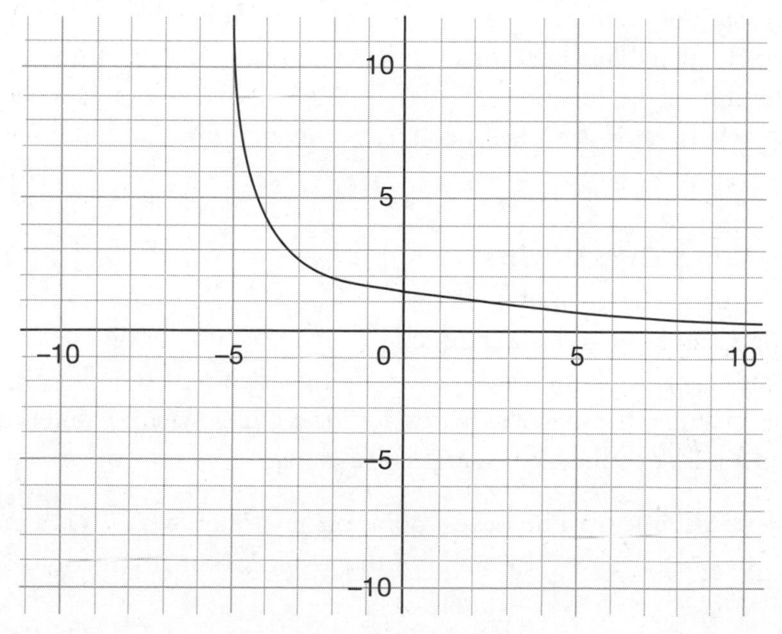

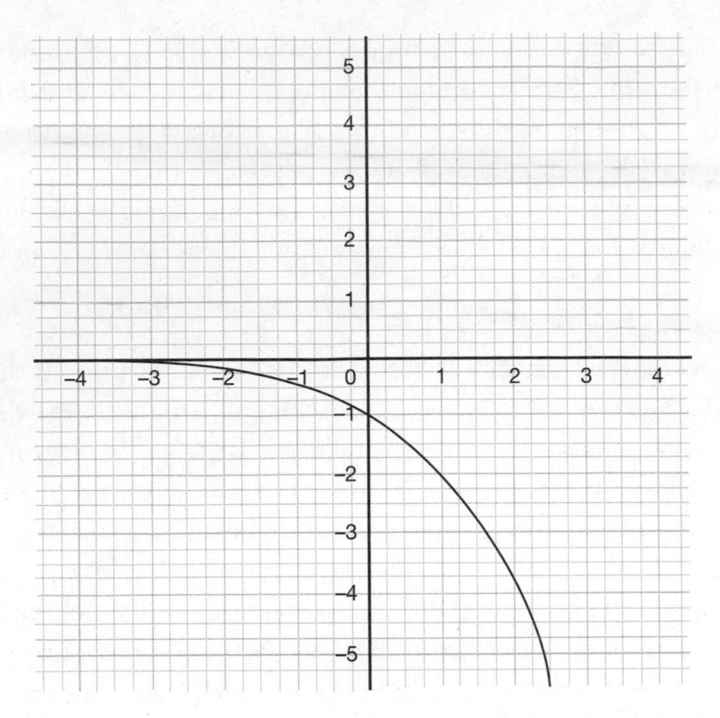

If you stated that the first graph illustrates a function that is decreasing at a decreasing rate then you were correct! That function starts out almost vertical when x is negative but soon after x turns positive, the slope mellows out to be almost a flat line with just a slight negative slope.

Hopefully you noticed that the second graph depicts a function that is decreasing at an increasing rate. Although the negative slope is barely perceptible at the left edge of the graph, the rate of decrease speeds up considerably as the function crosses the y-axis. Not long after x becomes positive, the rate of decrease increases so quickly that the function dives off the graph.

Describing Quadratics

Now we are going to apply what we've learned about rates of change to describe quadratics and other functions. Aside from being able to accurately and specifically describe the rate of change of a function, it's also helpful to know the correct words for describing when a function has reached a peak or valley, or changes directions:

- **Minimum:** The lowest point on any function. This is a measure of the y-value and not the x-value. Not all functions have minimums. If a function has a minimum at the point (2,4), the height of the function at 4 is the minimum value, so this would be described as, *the function has a minimum of 4 at 2.*

- **Maximum:** The highest point on any function. This is a measure of the y-value and not the x-value. Not all functions have maximums. If a function has a maximum at the point (2,4), the height of the function at 4 is the maximum value, so this would be described as, *the function has a maximum of 4 at 2.*
- **Turning point:** A point on a graph where the slope changes directions from negative to positive or vice versa.

Example: Describe the behavior of the following function:

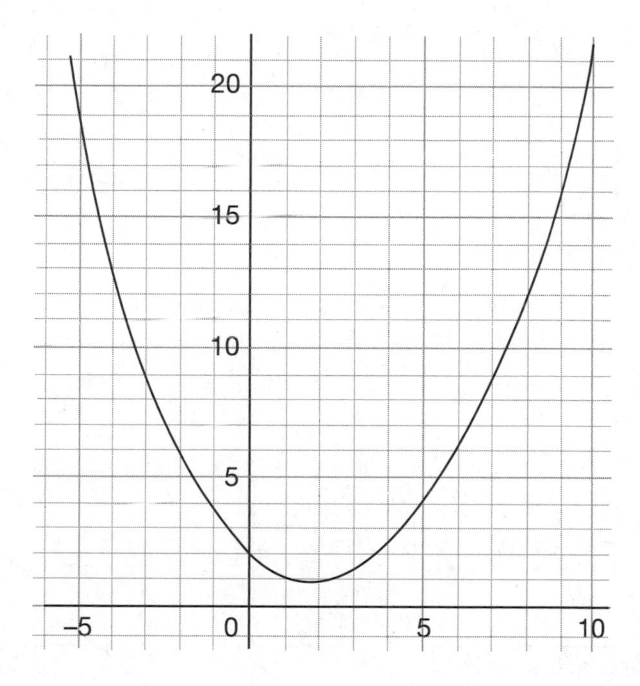

Example: Over the interval of $-\infty$ to -2, the function is decreasing at a high but decreasing rate. The rate of decrease slows down until the function reaches a minimum of 1 at $x = 2$. After this turning point the function begins to increase at a slow but increasing rate. After 5 this function has a very steep increase and it continues to increase at an increasing rate.

Practice 2

Use graphs A through F to answer the questions that follow.

A.

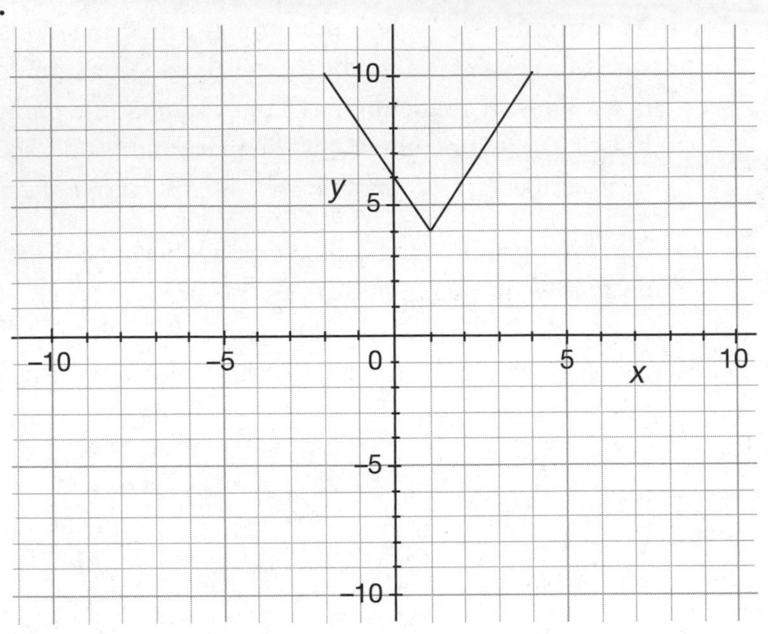

B.

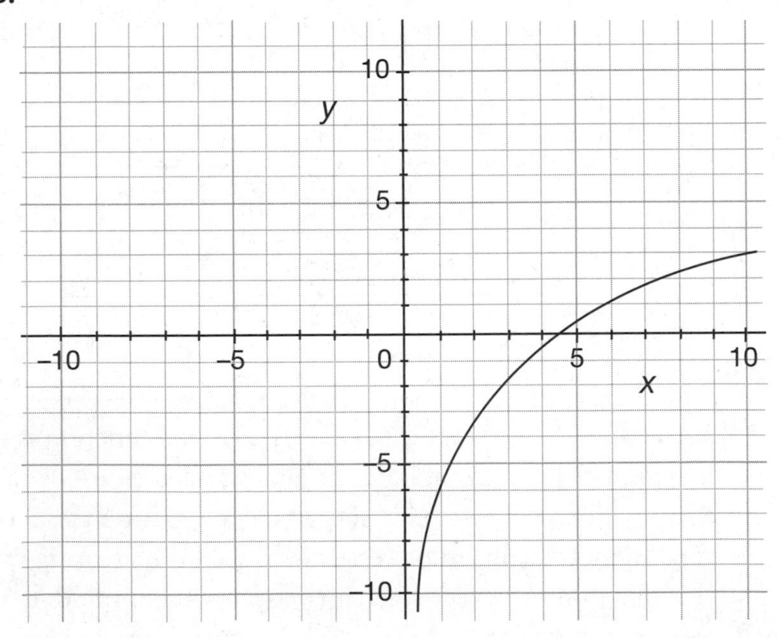

C.

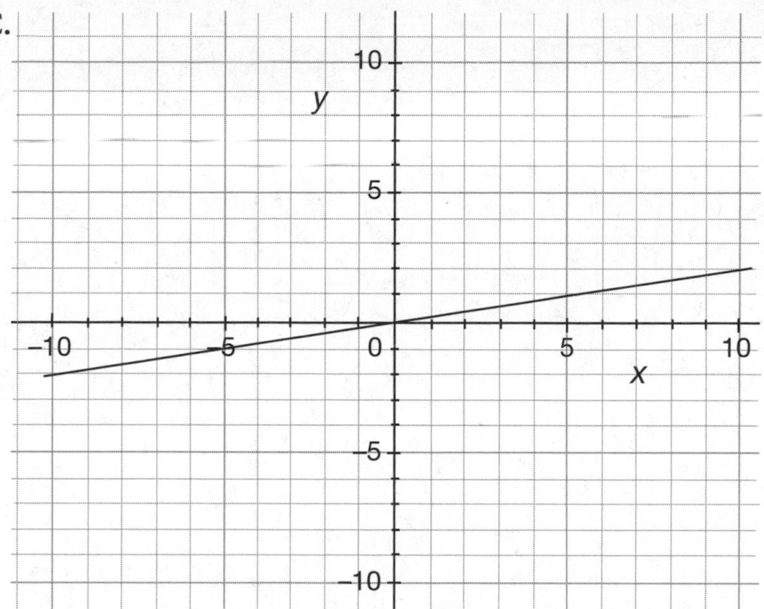

D.

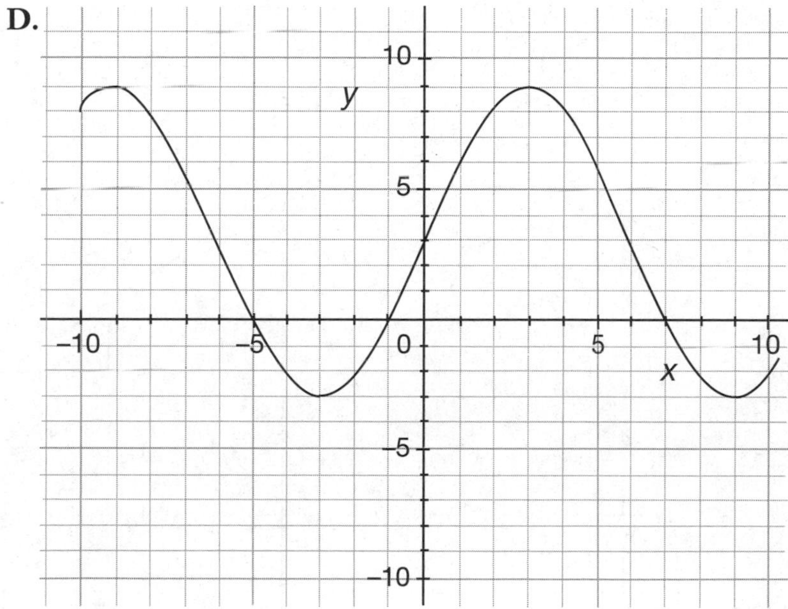

E.

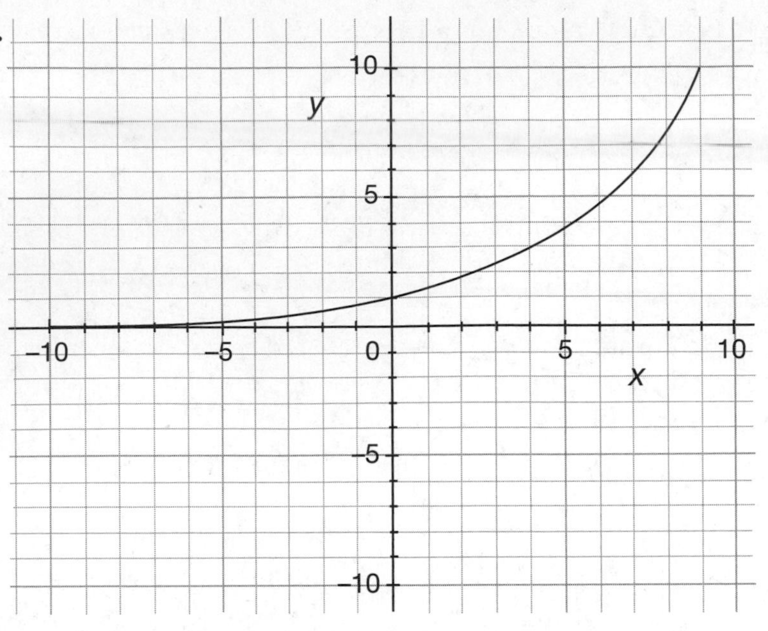

F.

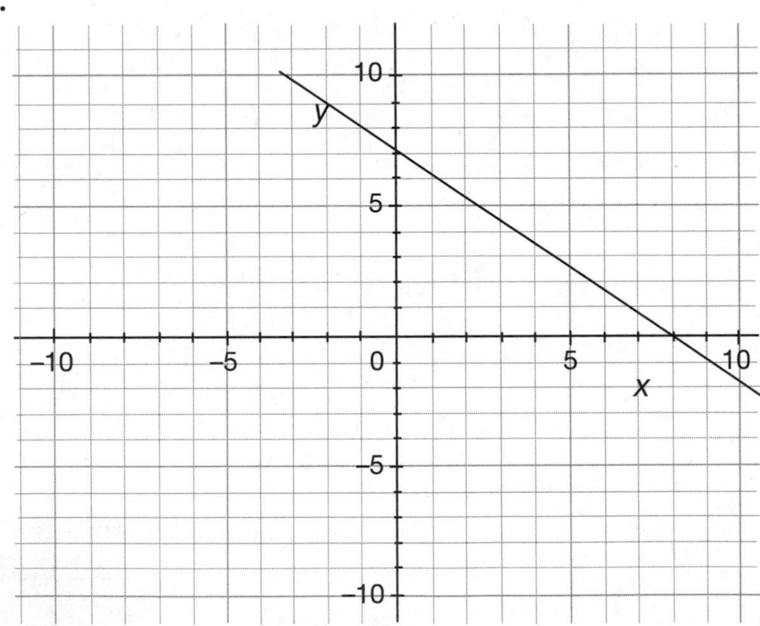

1. Which function(s) has/have a steady rate of change?

2. Which function(s) is/are increasing at an increasing rate?

3. Name the minimums and/or maximums of functions that have a minimum and/or maximum.

4. Which function(s) is/are non-linear?

5. Which function(s) has/have more than one x-intercept? What is significant about the x-intercept of a function?

6. How would you describe the rate of change for the function in graph B?

7. Which function could represent the temperature over the course of a year, over the interval $0 < x < 12$?

8. Do all functions have to have exactly one x-intercept and y-intercept? Explain your response by using the previous graphs to illustrate your response.

Drawing Functions from a Description

Now that you have a more refined eye for viewing slope, you will learn to sketch a slope from a verbal description. We will consider the way in which revenue increases over time for a new retail shop called Goods. Read each description and then look at the corresponding graph as you read the following story.

- Over the course of its first three months, Goods experiences a very moderately increasing revenue because it is just getting the word out to the community about its excellent offerings. (interval 0 to a)
- Once they have accumulated a base of interested customers, Goods' revenue begins to increase at a slightly increasing rate over two months as those customers start telling their friends about this great new business. (interval a to b)
- All of a sudden, a local blogger starts hyping up this hot new store and a buzz flies around town that Goods is *the* new spot to shop. For a month Goods sees a powerfully increasing rate of increase in their revenue. (interval b to c)

- After that busy month Goods is having trouble keeping up with demand. Their inventory runs low and customer service suffers a bit. So for two months the revenue of Goods is still increasing because they are the best shop around, but because of their dwindling inventory their revenue sees a decreased rate of growth. (interval *c* to *d*)
- For the next four months Goods' revenue has stabilized and is no longer increasing, but is not decreasing. They have a core group of loyal customers, but the fanatic shopping addicts are off checking out the next hot shop. (interval *d* to *e*)

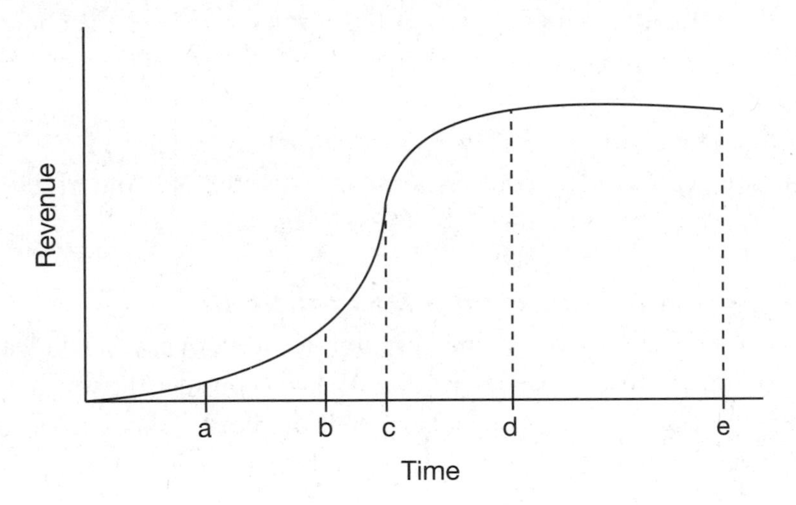

Now you will get to test your own graph-making skills in this culminating practice set.

Practice 3

Use the following information to answer the questions that follow. You will eventually be making a graph that compares her minutes of travel to the miles traveled.

Chi Chun leaves her home and walks two miles in 40 minutes to get to a bus stop. She just misses the bus and has to wait 20 minutes for the next bus. Finally it comes and she travels 38 miles over the course of 1 hour on the bus. Then she arrives at the train station and waits for 40 minutes before her train departs. The train comes and takes her 160 miles in just 90 minutes. She gets off the train and waits 20 minutes for her friend to arrive. Her friend arrives and it takes them 30 minutes to drive 30 miles to her friend's home. Chi Chun is exhausted from her long day of travel and needs a nap!

1. Use this information to fill in the following table. The first column should be intervals of time in minutes—there should be no gaps in time. The second column should show how many miles she traveled in each time interval, and the third column should show the total distance she has traveled since leaving the house. Use the fourth column to note what is happening with Chi Chun's travels. Some entries have been filled in for you:

Time Interval in Minutes	Distance during Interval	Total Distance	Notes
0–40			
	0 miles	2 miles	
			Bus to train
		40 miles	
160–250			Speedy train
	0 miles		
			Arrives at house & naps

2. What was her rate of speed in miles per hour for walking?

3. What was her rate of speed in miles per hour on the train?

4. What will the graph look like when Chi Chun is waiting for the bus, the train, and her friend?

5. What was her total number of hours traveled? Find her average rate in miles per hour.

6. Use the graph on the following page to construct a function that shows Chi Chun's travel over the course of her day. Use time in minutes as your independent variable and distance in miles as your independent variable.

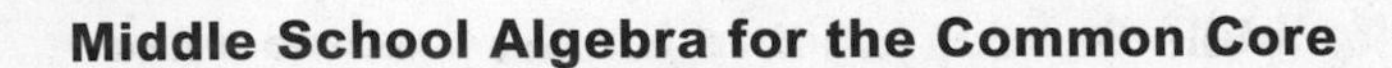

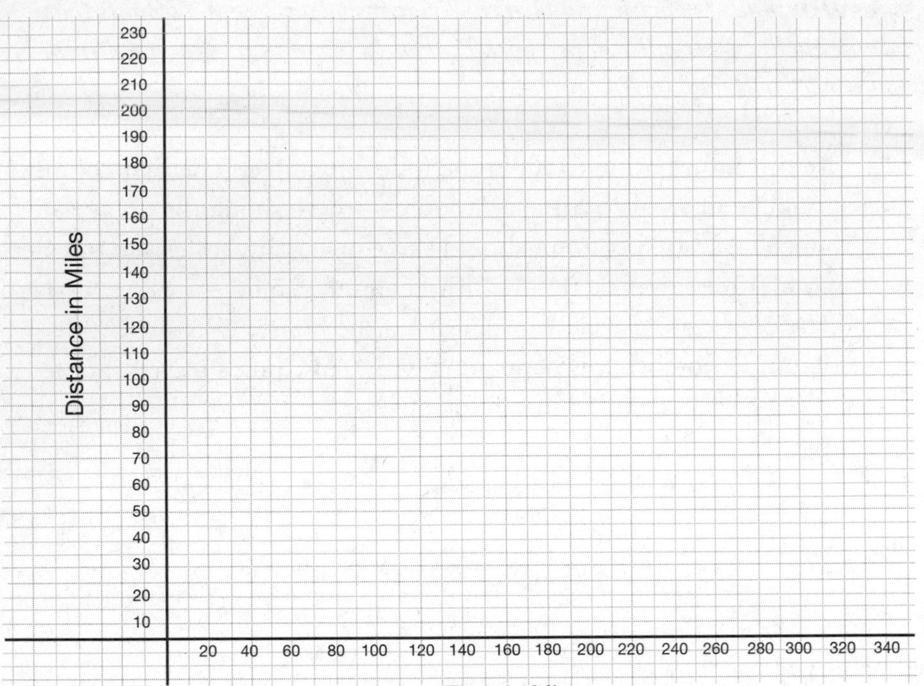

Distance in Miles
230
220
210
200
190
180
170
160
150
140
130
120
110
100
90
80
70
60
50
40
30
20
10
20 40 60 80 100 120 140 160 180 200 220 240 260 280 300 320 340
Time in Minutes

Answers

Practice 1

1.

x-Intervals	Function Behavior	Real-World Interpretation
0–30	Increasing at slowest rate	Warming up at rate of 15 minutes per mile
30–40	Slope of 0	Stop to stretch for 10 miles after her first two miles or maybe eat a power snack
40–52	Increasing at steepest rate	Run at almost sprint speed of six minutes per mile—that's so fast!
52–70	Slope of 0	Stop to cool down and stretch and enjoy the nice day
70–91	Decreasing at steep rate	Fast run back toward home—when one mile from home she notices that her friend is home and stops to spend time with her

2. Sierra's rate of change during her first two miles is $\frac{15}{1}$, meaning that she ran one mile in 15 minutes.
3. Sierra's distance from home remained at two miles for 10 minutes, so it appears that she stopped running during this time.
4. Sierra's rate of change from 40 to 52 minutes was $\frac{6}{1}$, meaning that she ran one mile in just six minutes—she was going more than twice as fast as she was going during the first 30 minutes.
5. The negative slope shows that she is headed back toward home since her distance from home is decreasing. Sienna is definitely not slowing down since her rate of change was three miles in 21 minutes, which is $\frac{7}{1}$ or seven minutes per every mile.
6. At 91 minutes Sierra was one mile from home. This graph does not show her running all the way back home, so she must have stopped her run. In total she ran seven miles.
7. See the table above.

Practice 2

1. Functions C and F have a steady rate of change. They are linear.
2. Function E is increasing at an increasing rate.
3. Function A has a minimum and Function D has multiple maximums and minimums.
4. Functions A, B, D, and E are non-linear.
5. Functions A and D have more than one x-intercept. The x-intercept is the break-even point of a function and represents where the value of y is 0.

6. Function B has a decreasing rate of increase.
7. Function D could represent the temperature fluctuations over the course of a year.
8. Not all functions will have an x-intercept or a y-intercept. Functions can have more than one x-intercept but can never have more than one y-intercept. Function A has two x-intercepts, but just one y-intercept. Function B has one x-intercept but appears to not have a y-intercept. Function D has multiple x-intercepts, but just one y-intercept. Function E has one y-intercept but appears to not have an x-intercept.

Practice 3

1.

Time Interval in Minutes	Distance during Interval	Total Distance	Notes
0–40	2 miles	2 miles	Walked
40–60	0 miles	2 miles	Waiting for bus
60–120	38 miles	40 miles	Bus to train
120–160	0 miles	40 miles	Waiting for train
160–250	160 miles	200 miles	Speedy train
250–270	0 miles	200 miles	Waiting for friend
270–300	30 miles	230 miles	Arrives at house & naps

2. $\frac{2}{\frac{2}{3}}$ = three miles per hour.
3. $\frac{90}{1.5}$ = 60 miles per hour.
4. During the times she is waiting, the graph will be flat.
5. She traveled for five hours. $\frac{230}{5}$ = 46 miles per hour.
6.

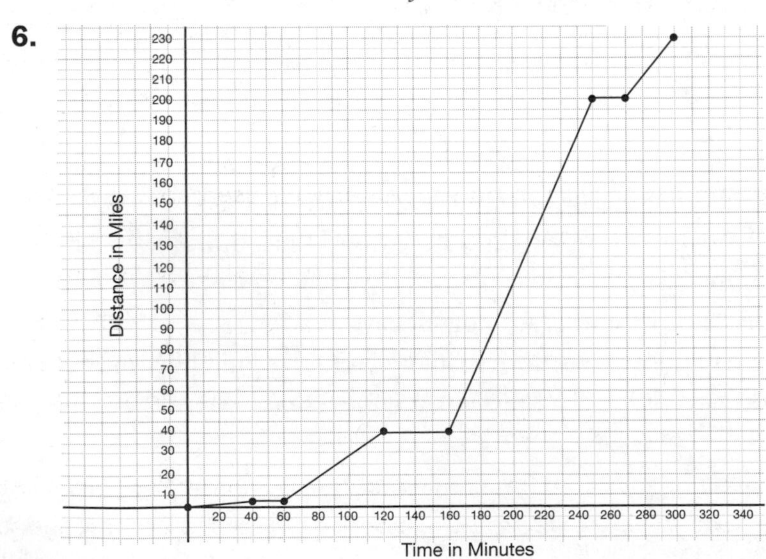

Additional Online Practice

Using the codes below, you'll be able to log in and access additional online practice materials!

Your free online practice access codes are:
FVE4I2I0O2NDO75BT277
FVEEP615DFNXW362R66Q
FVEIGT3751V6TO263QP6
FVE525J3587GUDHM165I

Follow these simple steps to redeem your codes:

- Go to **www.learningexpresshub.com/affiliate** and have your access codes handy.

If you're a new user:

- Click the **New user? Register here** button and complete the registration form to create your account and access your products.
- Be sure to enter your unique access codes only once. If you have multiple access codes, you can enter them all—just use a comma to separate each code.
- The next time you visit, simply click the **Returning user? Sign in** button and enter your username and password.
- Do not re-enter previously redeemed access codes. Any products you previously accessed are saved in the **My Account** section on the site. Entering a previously redeemed access code will result in an error message.

If you're a returning user:

- Click the **Returning user? Sign in** button, enter your username and password, and click **Sign In**.
- You will automatically be brought to the **My Account** page to access your products.
- Do not re-enter previously redeemed access codes. Any products you previously accessed are saved in the **My Account** section on the site. Entering a previously redeemed access code will result in an error message.

If you're a returning user with new access codes:

- Click the **Returning user? Sign in** button, enter your username, password, and new access codes, and click **Sign In**.
- If you have multiple access codes, you can enter them all—just use a comma to separate each code.
- Do not re-enter previously redeemed access codes. Any products you previously accessed are saved in the **My Account** section on the site. Entering a previously redeemed access code will result in an error message.

If you have any questions, please contact LearningExpress Customer Support at LXHub@LearningExpressHub.com. All inquiries will be responded to within a 24-hour period during our normal business hours: 9:00 A.M.–5:00 P.M. Eastern Time. Thank you!

NOTES

NOTES

NOTES

NOTES